# Biochemistry
## Laboratory Manual
### BIOL 3612

Editor
Kirsten Fertuck, Ph.D

Biology Department
Northeastern University

**Biochemistry Laboratory Manual – BIOL 3612**
Copyright © 2017 by Kirsten Fertuck, Biology Department, Northeastern University



Printed in the United States of America

ISBN-13: 978-1-68284-229-4
ISBN-10: 1-68284-229-0

# BIOCHEMISTRY

# BIOL 3612
# Laboratory Manual

# Table of Contents

# Essential Information



## Laboratory Manager:

Frauke Argyros  f.argyros@northeastern.edu
703 Behrakis   617-373-2118

The laboratory manager should be your second point of contact regarding the graded aspects of this course, if your inquiries are unresolved after speaking with the TAs. The laboratory manager also handles academic dishonesty cases – **avoid** being involved in any such problem by conscientiously doing your **own** work, and **always** acknowledging others in writing if you use their ideas or work!

## Teaching Assistants:

Contact information and office hours will be provided. Please note that you are welcome to go to any TA's office hours based on your scheduling preferences – you aren't restricted to visiting your own. The TAs should be your first point of contact if you have questions regarding graded aspects of this course. Please be aware that TAs are not expected to be 'on call' at all times for this course; you should make note of and respect their stated office hours, during which time they will be happy to assist you!



Maintaining a safe lab environment is naturally of paramount importance, and will allow us to focus on our learning objectives for the semester. You should expect that the hazards of any materials that we use will be made clear when applicable; nonetheless, it is your responsibility to behave in a responsible manner within the laboratory environment. This includes strict adherence to the following:

- No eating, drinking, chewing gum, or applying cosmetics
- No open-toed shoes
- Wearing protective equipment as directed
- Being respectful and following the instructions of the TA and lab manager at all times
- Maintaining a clean work environment and disposing of all waste in the manner directed
- Labeling all samples clearly and storing them properly

Maintaining an **accurate, honest, and complete record of your laboratory work** is an essential component of any laboratory experience – whether as a part of your coursework, a volunteer or paid internship, or in a more long-term professional setting.

A laboratory notebook is a critical record of work that has been done, allowing the experimenter or another reader to **fully recreate the work that has been done and to clearly view and understand all results that have been obtained**.

Each day's experiment **must** begin with the following:
- Date (including year)
- Name and name of lab partner(s)
- Experiment number and title
- Brief description of purpose

Next, be certain to include the following:
- All experimental details that would allow someone to replicate the experiment – be **clear and exhaustive** here! Details (temperatures, volumes, voltages, types of equipment, enzyme expiration dates, etc.) are key to data interpretation as well as the ability to later reproduce results
- All experimental results: **both** qualitative observations (e.g. color change, formation of precipitate) **and** raw quantitative data (be certain to include correct units and the appropriate number of significant digits for the type of data)
- All images, or detailed information regarding where image files are stored (e.g. sketch an absorption spectrum **and** indicate the filename <u>and</u> exactly where the file is stored)
- All data analyses, including calculations
- Data interpretation, including possible sources of error

And finally, this should be implied above, but it bears repeating: you should **never** assume that you would have the opportunity to communicate with someone as they look through your lab notebook and prepare to replicate your work – your notebook should be a **standalone** document that **already contains the complete record**. Not only is expected that a lab course provides training in this, but creating this stellar documentation should be a source of pride!

It is essential that you understand that we use a **zero tolerance policy on academic dishonesty**. Scientific advancements involve collaboration and building on the work of others, and it is essential to credit the work of others that allows you to make new observations and inferences.

Course credit can never be given for plagiarized work, and of course positive reference letters also can not be written for students who plagiarize the work of others!

**Avoid problems by always citing your sources of information, crediting students who gave you data (images or otherwise) to use, and creating original responses to every written assignment.**

# Lab grading scheme

| | |
|---|---|
| 25pt | Maximum of 5x5pt for the copy of the lab notebook pages that are handed in at the end of labs 1A, 2A, 3A, 4A, and 5A. The proper type of lab notebook must be used (as specified in the bookstore listing for this course) that creates a duplicate record of each of your note pages. For full credit, lab notes must be legible, dated, and consist of a complete record of experimental procedures and results that allow a reader to understand what work was performed, everything that was observed and measured, and what data were obtained from other people. |
| 50pt | 5x10pt quizzes at the start of labs 1A, 2A, 3A, 4A, and 5A based on assigned readings. The quizzes are intended to emphasize the importance of coming on time and coming prepared to lab. |
| 55pt | DNA Isolation and Analysis report. Contains complete sentences; answers all questions; includes, properly labels, and discusses all requested results and figures in proper format from Lab 1A (DNA isolation) and the agarose gel from Lab 2A. |
| 60pt | Enzyme Kinetics report. Contains complete sentences; answers all questions; includes, properly labels, and discusses all requested results and figures in proper format from Lab 3A (enzyme kinetics) and Lab 4A (visualization of an enzyme active site). |
| 60pt | Protein Concentration and SDS-PAGE report. Contains complete sentences; answers all questions; includes, properly labels, and discusses all requested results and figures in proper format from Lab 2A (protein concentration by Lowry method) and Lab 5A (SDS-PAGE and protein concentration by A280 method). |
| 50pt | 5x10pt assignments in labs 1B, 2B, 3B, 4B, and 5B. |
| **300pt** | **Total number of points available, which will be converted to 100pt and then to a letter grade as shown in the table below.** |

| | |
|---|---|
| A | 93-100 |
| A- | 90-92.9 |
| B+ | 87-89.9 |
| B | 83-86.9 |
| B- | 80-82.9 |
| C+ | 77-79.9 |
| C | 73-76.9 |
| C- | 70-72.9 |
| D+ | 67-69.9 |
| D | 63-66.9 |
| D- | 60-62.9 |
| F | < 60 |

## Quizzes

There are no make-up quizzes. The lab course requires that you be on time and prepared for the work that will be performed that day, and having a quiz at the beginning of the lab period is designed to help with both of those objectives.

## Assignments

**Timely submission of work should be a priority** for this lab. There are eight submission deadlines (for three major assignments and five dry lab assignments). **It is your responsibility to make sure that you know and adhere to these eight assignment deadlines!**

The three major assignments of the semester (DNA Isolation and Analysis, Enzyme Kinetics, and Protein Concentration and SDS-PAGE) are worth 55-60pt and have late penalties of 10pt per day, resulting in a loss of ~17-18% per day late. **Avoid this** by turning in your work well in advance of the deadline, and always saving your Turnitin receipt as confirmation that you submitted your assignment on time.

The remaining five assignments (for dry labs 1B-5B) are worth 10pt each, and have late penalties of 5pt per day, resulting in a loss of 50% after the first day and no credit after the second day. **Avoid this** by turning in your work well in advance of the deadline. These dry labs are intended to be **completed during your assigned lab period, including the submission of the assignment itself**, so avoid waiting until the formal deadline! As always, save your Turnitin receipt as confirmation that you submitted your assignment on time.

# Course overview

The lab portion of the biochemistry course consists of five wet labs (Labs 1A-5A) that take place in the biochemistry lab facility and five dry labs (Labs 1B-5B) that you perform on your own using the lab materials for guidance.

The overall purpose of the lab is to complement the lecture by providing hands-on exposure to selected techniques and analyses that are difficult to appreciate in a lecture setting alone. Wet labs focus on working with nucleic acids and proteins, using modern instrumentation and techniques, reproducibly handling small volumes, and practicing proper procedure for data recording and analysis. Dry labs focus on freely-available tools that are immensely useful to biochemists, including but not limited to a visualization program that is used to make 3D representations of structures that are challenging to understand in flat textbook representations.

In order to provide some continuity, student-isolated DNA as well as four different proteins are examined in several different labs as follows:

| Lab name | Lab description | Strawberry DNA | Albumin | Alkaline phosphatase | Chymotrypsin | Malate deHase |
|---|---|---|---|---|---|---|
| Lab 1A | DNA isolation by two methods | X | | | | |
| Lab 1B | Genome database exploration and PCR primer selection | | | X | | |
| Lab 2A | Agarose gel of isolated DNA; protein colorimetric assay | X | X | | | |
| Lab 2B | Introduction to Jmol and protein calculator tools | | | | | |
| Lab 3A | Kinetic study of alkaline phosphatase | | | X | | |
| Lab 3B | Creating favorable conditions for protein synthesis | | | | | |
| Lab 4A | Alkaline phosphatase kinetics and active site | | | X | | |
| Lab 4B | Enzyme active site residues and substrate binding; conversion of a zymogen to an active enzyme | | | | X | |
| Lab 5A | SDS-PAGE and protein conc. by A280 method | X | X | X | X | X |
| Lab 5B | Protein disulfide bonds, glycosylation, and redox reactions | | X | X | X | X |

# Lab 1A: DNA isolation by two methods

Overview of Lab 1A

- Introduce overall goals and course policies.
- Take first quiz: spectrophotometry of nucleic acids.
- Review some basic lab techniques: filtration, pipetting, centrifugation, spectrophotometry.
- Isolate chromosomal DNA by two different methods, and compare yield and purity.

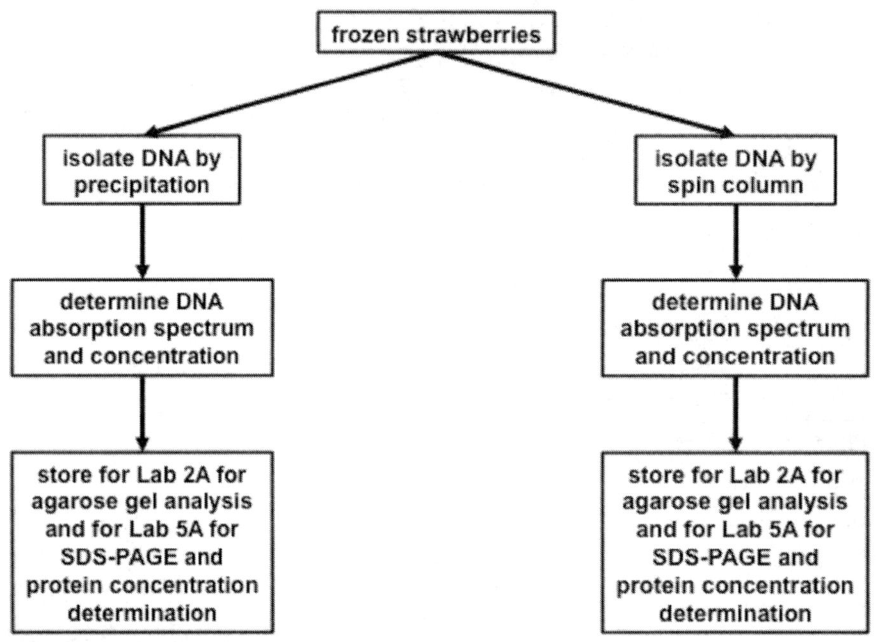

- Complete first quiz (**Quiz 1A**) on assigned reading: spectrophotometry reading from **NanoDrop Nucleic Acid Handbook (p.11-12)**.
- Complete course introduction:
    o   TA and lab coordinator contact information; TA office hours
    o   Lab safety regulations and procedures
    o   Procedures for submitting assignments; clarification of the eight assignment deadlines
    o   Academic honesty policies
- Isolate DNA from strawberries by precipitation method and by Qiagen DNeasy Plant Mini column.
- Become familiar with the proper use of NanoDrop spectrophotometers and the extent to which they provide reproducible data. These instruments will be used in all five wet labs (1A-5A).
- Examine and save full absorbance spectra of DNA samples using NanoDrop spectrophotometer; compare spectra obtained by the two DNA extraction methods.
- Record DNA concentration and calculate the mass of DNA that will be loaded in the agarose gel for Lab 2A (i.e. mass of DNA in a 13.5ul volume of DNA sample)
- Calculate yield of DNA (mass) compared with original mass of strawberry material.
- Carefully label and freeze DNA samples for use in future labs.


Preparation for Lab 1A

- Prepare for **Quiz1A** with assigned reading from the **NanoDrop Nucleic Acid Handbook (p.11-12; provided on Blackboard)**.
- Read over the complete procedure for the lab (including both isolation methods).

1. **Become familiar with the use of the NanoDrop 2000c spectrophotometer.** As a class, watch the video on the use of the NanoDrop instrument https://www.youtube.com/watch?v=FiGZnNs2xXY (the first part is about DNA; skip the second part, which is about fluorescence). **Note that these instruments are very delicate, and always ask if you need further instruction!**

2. **In consultation with your TA, determine whether you will be isolating strawberry DNA by the precipitation method or the column method.** One half of the class will perform each method. Detailed procedures follow below for each method.

3. **Start your laboratory notebook record by indicating the procedure that you will be following and the partner(s) that you will be working with.** As you go along, be certain to write down **all steps and observations** carefully in your notebook.

   o As you go, make particular note of any deviations from the written procedure, whether accidental (e.g. due to a spill) or intentional (e.g. a purposeful change in the amount of starting material that your TA decides that you will use).

   o **Remember that a copy will be handed in for your TA to grade; it must be legible and must have sufficient information to allow someone to understand your observations <u>and</u> reproduce your work! Do not try to save paper and squeeze all of your observations into a small space – make them neat and easy to read!**

4. **Isolate the strawberry DNA by your assigned method.** Consult the relevant procedure (precipitation or column) that follows.

5. **Determine the DNA concentration in your final purified DNA sample.** After isolating the DNA by your assigned method, determine the DNA concentration of your sample using the NanoDrop spectrophotometer, **asking for guidance as needed with this delicate instrument**! For aqueous solutions such as this DNA solution, always carefully measure and dispense **1ul** with a 10ul pipettor, using a properly-fitting pipet tip. After the measurement, carefully wipe the solution from the top and bottom of the NanoDrop pedestal using a lint-free lab wipe – you must do this diligently after **every** measurement so that samples do not dry onto the delicate instrumentation!

   o Make a table in your notebook like the one that follows, with five columns where you will record the absorbance values at the two key wavelengths (**A260 and A280**) as well as the two key ratios (**A260/A230 and A260/A280**) and the **concentration (ng/ul)**. Leave enough rows so that you can measure your sample in triplicate if time permits, to help you to establish the reproducibility in your measurements.

| measurement | A260 (A.U.) | A280 (A.U.) | A260/280 | A260/A230 | concentration (ng/ul) |
|---|---|---|---|---|---|
| replicate 1 | | | | | |
| replicate 2 (if time permits) | | | | | |
| replicate 3 (if time permits) | | | | | |

- o  It is essential to first 'blank' the spectrophotometer using 1ul of solvent (water in this case) **that your DNA sample is dissolved in** – this informs the instrument about any absorbance at any wavelength that is attributable purely to the solvent. There is nothing to record here.

- o  After wiping off the 'blank' with a lint-free lab tissue, measure the absorbance of 1ul of your DNA solution. **Fill in all five entries** in the first row of your table. If there are any error messages (indicating that the software suspects contamination or other problems in your sample), be sure to note them. This may appear, for example as a blue lower case 'i' symbol next to the reading.

  - ▪  (Note: as you should recall from the modified Beer-Lambert equation on p.12 of the **NanoDrop Nucleic Acid Handbook**, the measurement is **volume-independent** – spectrophotometers measure absorbance through a fixed path length through a sample. As little as 1ul is fine for DNA-containing samples, while 2ul is recommended for protein-containing samples due to how the detergents that are often found in protein solutions affect the properties of water and specifically its ability to form the needed column between the top and bottom of the NanoDrop. If you have time, you can try to verify experimentally that (e.g.) 1.0ul, 1.5ul, and 2.0ul of your DNA sample all yield the same absorption data.)

- o  Make a brief sketch of the absorption spectrum. **You shouldn't spend more than two minutes on this.** Quickly note the range of wavelengths being monitored on the x-axis, sketch the overall pattern of the absorption curve, and note the approximate wavelength of any 1-2 peaks that have clearly higher absorption than anything else on the spectrum. (If the spectrum is 'noisy' with a large number of peaks, then don't note their individual wavelengths.) If desired/needed, you can take a photo of the screen and refer to it while making your sketch back at your bench, so that someone else can start using the spectrophotometer. One sketch is sufficient, even if you make triplicate readings.

- Remember, p.11 of the **NanoDrop Nucleic Acid Handbook** shows a 'typical' absorption spectrum for a pure nucleic acid sample. Pages 18-19 show examples of spectra for that are indicative of a variety of different problems, either with sample contamination or with the instrument.

6. **Share data with a group that performed the other DNA isolation method.** If time permits, share data with a group that performed the other isolation method. (If not enough time remains, then take some quick photos of the other group's notes, or make other arrangements, so that you will still have this information for your lab report.) Write down the following information in your notebook from a group that did the **other method** of DNA isolation:

   o The names of those students
   o The name of the method that they used
   o The mass of strawberry that they used
   o Their five columns of data from the spectrophotometer (one row is fine – you don't need triplicate data)
   o Any notes on intentional or unintentional deviations from the procedure that they made
   o A quick sketch of their absorption spectrum, noting whether or not peak absorbance was attained at A260

7. **Clean up**. Spectrophotometers should be turned off (switch is at the back), and the upper and lower parts of the pedestal (the parts that came in contact with the DNA solution) should be carefully and thoroughly wiped with a water-soaked lint-free tissue and then a dry tissue, so that no DNA dries on. Clean up your bench completely, putting all dirty glassware and plasticware in the appropriate receptacles.

8. **Look over the assignment questions.** If you have any remaining time after your notebook is complete, look over the questions for the DNA Isolation and Analysis assignment.

9. **Hand in a copy of your lab notebook pages for grading before you leave at the end of the lab period**.

- Remember, these may **actually be used** in order to see what you personally did and observed, in order to work on further refining the labs – it is **essential** that your notes be complete, legible, and focused on what you personally did and found, not just the original typed-out plan for how the lab could have gone. **Do not try to conserve notebook space!** Take as much room as you need in order to preserve a clear record of your actions and observations.

- In addition, keep in mind that the work that you are performing today will be used in an assignment ('DNA Isolation and Analysis') that will be due later, after you perform additional work in Lab 2A. This is very typical of actual lab work, where notes and data are frequently referred to months or years after they were originally made/obtained. Therefore, this clear and complete record is needed for your own use, in addition to being for the TA and the course organizers!

This method for isolating DNA from strawberries focuses on starting with a large amount of strawberry tissue in order to obtain a visible precipitate in the later steps. It was the only DNA isolation method in the biochemistry laboratory manual for many years, adapted from a protocol from Rebeca Rosengaus, Ph.D; in this lab we will compare this procedure with a proprietary column-based DNA purification method.

## Materials:

> frozen strawberries
> previously-prepared salt/detergent solution (7.5g NaCl, 50ml Joy detergent, 1L distilled water), warmed to 60C in preparation for the experiment
> ice-cold ethanol, stored on ice during the experiment
> ice bucket
> test tube
> cheesecloth (or coffee filter)
> 2 beakers (250ml)
> rubber bands
> mortar and pestle
> hot plate
> wooden sticks
> glass hooks (can be made with Pasteur pipets)
> Pasteur pipets and bulbs for transferring ethanol

## Procedure:

1. Weigh 2 fresh strawberries and record their mass. **Be careful to record the correct number of significant digits!**

2. Place the strawberries in a mortar and mash them to a pulp with the pestle, breaking as much solid material as possible.

3. Transfer the pulp to a 250ml beaker and add an approximately equal amount of heated saltwater/detergent solution (9-12 eyedrops). Stir to create a homogenous suspension.

4. Transfer the mixture to a 250ml beaker that has been previously covered with cheesecloth, secured with a rubber band. Using a wooden stick, separate the pulp over the cheesecloth as much as possible, encouraging the liquid to drip through (~5min).

5. Discard the pulp-containing cheesecloth in the trash, and transfer the filtrate to a test tube so that the tube is only **half**-filled. Be sure to record approximately how much remains.

6. Tilt the test tube at a 45-degree angle and **very slowly** start adding the ice-cold ethanol so that it drops down the test tube wall. **Do not shake the test tube – the ethanol should form a layer on top of the filtrate**. Add enough ethanol to fill the test tube 2/3 of the way. One layer should now be red and the other clear.

7. Look closely at the interface where the ethanol and the red layer make contact. You should start seeing some white, bubbly material which becomes 'stringy' and very viscous – DNA that has precipitated from the strawberry filtrate! Use a glass hook and introduce it very slowly to that interface. Swirl the glass hook just at this level. DNA will start clumping further and you can lift this genetic material right out of the test tube and transfer it to a new 1.5ml tube. Unlike some methods, with this isolation method you are actually getting to see the DNA itself!

8. Air-dry the DNA briefly (~2-5min) to allow the ethanol to evaporate, and, with your TA's approval, resuspend it in 500ul of water. (You will likely see that it does not fully dissolve, since this brief isolation procedure does not contain wash steps to try to remove contaminating reagents that were used in the isolation.)

9. Label the tube clearly with your initials and section number so that you can find it again in future labs.

This method for isolating DNA from strawberries uses a widely-used type of 'spin column' that is made by the company Qiagen. It uses materials that are sold as part of the "DNeasy Plant Mini kit" and the procedure below is adapted from the accompanying handbook. Note that overloading the "mini" column could result in low purification success, and therefore a small amount of strawberry material is used as the starting material – **unlike** in the precipitation method, where our primary goal is to have enough DNA to be able to physically observe it precipitating from solution.

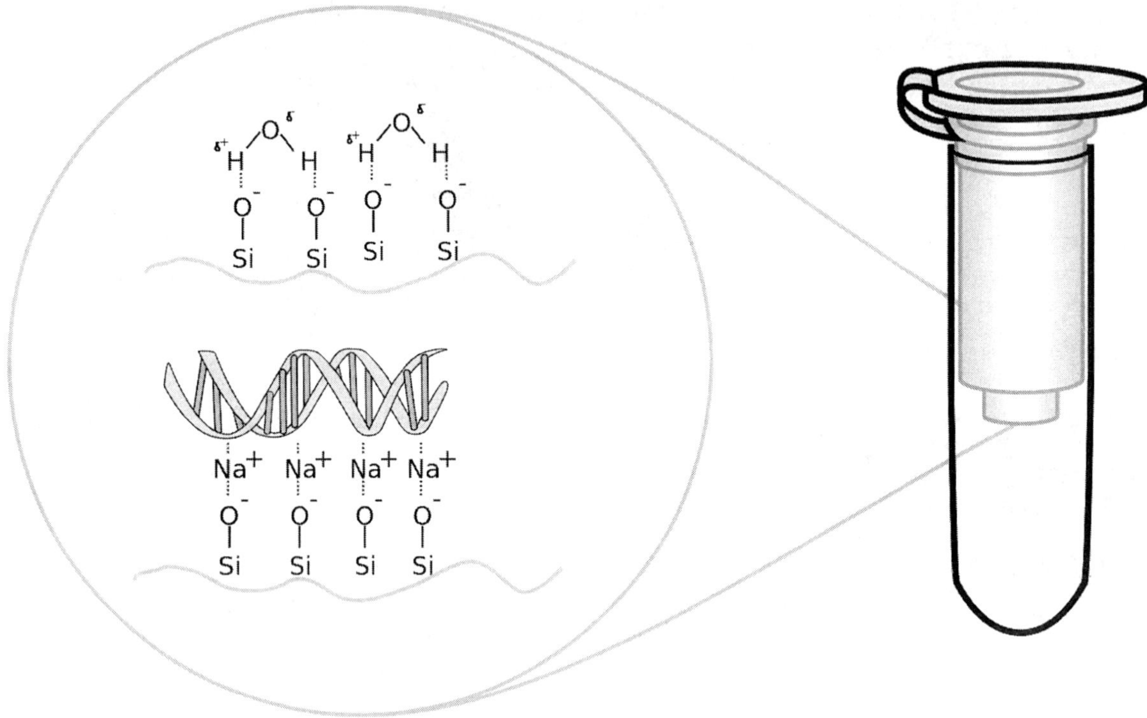

This figure, available in the public domain at https://commons.wikimedia.org/wiki/File:Qiagen_Mini_Spin_Column.svg shows a spin column and a rendering of the binding surface. The exact chemistry is proprietary, but the company describes it as a "silica-based membrane". Changing the chemical conditions in which the DNA is suspended affects whether it will bind to the membrane (required for the early binding and washing steps) or not (required for the final 'elution' step, in which the DNA will be collected in a tube below).

**Materials:**

        approximately 0.5g fresh strawberry
        heat block or water bath heated to 65C
        ice bucket and ice
        Buffer AP1 (proprietary lysis buffer)
        RNAse A
        Buffer P3 (proprietary buffer to precipitate detergent, proteins, and
            polysaccharides)
        Buffer AW1 (proprietary wash buffer)
        Buffer AW2 (proprietary wash buffer)
        water (solvent for DNA sample)
        QIAshredder spin column
        DNeasy spin column
        2ml centrifuge tubes

**Procedure:**

1. Obtain a piece of strawberry that is approximately 0.5g (approximately 0.5 cm$^3$) and record its mass. **Be careful to record the correct number of significant digits!**

2. Pulverize the strawberry as directed. **This is a critical step** – there is room for creativity here in how you mash/macerate the strawberry, but achieving a fine pulp here is crucial to a high yield, so keep working until no chunks remain. Finely chopping/mashing the tissue with a metal blade in a weighing dish can work well.

3. Carefully transfer the mashed tissue to a 1.5ml tube. Add 400ul Buffer AP1 and then 4ul RNAse A. Vortex and incubate 10min at 65C. Invert tube 2-3x during incubation.

4. Add 130ul Buffer P3. Mix and incubate 5min on ice.

5. Centrifuge lysate for 5min at 20,000 x g (14,000 rpm).

6. Pipet lysate into a QIAshredder spin column placed in a 2ml collection tube. **(Check the package label – make sure you are using a QIAshredder here!)**

7. Centrifuge for 2min at 20,000 x g.

8. Transfer the flow-through liquid into a new 1.5ml tube, avoiding any solid pellet that may be present. Estimate the volume of liquid, and whatever it is (X), add 1.5X of Buffer AW1 (e.g. if the flow-through is ~500ul, add 750ul of Buffer AW1). Mix by pipetting.

9.  Transfer 650ul of the mixture into a DNeasy Mini spin column placed in a 2ml collection tube. (**Check the package label – make sure you are using a DNeasy Mini spin column here!**)

10. Centrifuge for 1min at ≥6000 x g (≥8000 rpm). Discard the flow-through. Repeat this step with the remaining sample.

11. Place the spin column into a new 2ml collection tube. Add 500ul Buffer AW2, and centrifuge for 1min at ≥6000 x g. Discard the flow-through.

12. Add another 500ul Buffer AW2. Centrifuge for 2min at 20,000 x g. **Note: remove the spin column from the collection tube carefully so that the column does not come into contact with the flow-through.**

13. Transfer the spin column to a new 1.5ml or 2ml tube.

14. Add 50ul water for elution. Incubate for 3min at room temperature. Centrifuge for 1min at ≥6000 x g.

15. Repeat the previous step without changing tubes, so that the final DNA solution after the second elution has a volume of 100ul.

16. Label the tube clearly with your initials and section number so that you can find it again in future labs.

# Lab 1B: Genome database exploration and PCR primer selection



- Review/examine three Nobel Prize-winning discoveries related to DNA structure and manipulation:
    - **DNA structure** – Nobel Prize in Physiology or Medicine, 1962; more information available at https://www.nobelprize.org/nobel_prizes/medicine/laureates/1962/perspectives.html
    - **Restriction enzymes** – Nobel Prize in Physiology or Medicine, 1978; more information available at http://www.nobelprize.org/nobel_prizes/medicine/laureates/1978/press.html
    - **Polymerase Chain Reaction (PCR)** – Nobel Prize in Chemistry, 1993; more information available at http://www.nobelprize.org/nobel_prizes/chemistry/laureates/1993/press.html
- Use the UCSC Genome Browser as a starting point to find relevant information about a gene and the structure and function of its protein product(s); explore some tracks related to gene expression; see how to view all of the restriction enzyme cut sites in a genome.
- Use a PCR primer-selection tool and correctly interpret results, including precisely where (both with respect to strand and orientation) primers will anneal; mark where in the Genome Browser the primers will anneal.



- **Watch an animation of DNA packaging and replication**, paying particular attention to the speed of template-directed synthesis and the depiction of the major and minor grooves.
- **Use the USCS Genome Browser** to explore the intestinal alkaline phosphatase gene in detail (which will then be further examined in Labs 3A and 4A).
    - Choose a species, search for a gene, and select specified tracks. View further information about the gene and protein.
    - View all of the predicted cut sites in the genome for a desired restriction enzyme.
- **Use Primer3 to design primers** that would amplify a portion of the intestinal alkaline phosphatase gene.
    - Identify how the primer output sequence will align with the input DNA sequence that is to be amplified, focusing on the directionality of the strands and which primer will bind to which strand.

o   Use the In-Silico PCR tool to show where in the Genome Browser these
    particular primers will anneal, and verify that only one amplicon is
    expected.


## Preparation for Lab 1B

The Genetics prerequisite for this course is intended to have provided sufficient
background on the topics of restriction enzymes and PCR. If needed, please review the
following resources to familiarize yourself with some of the ways that these tools are
commonly applied:

- **Restriction enzymes**: http://www.jove.com/science-
  education/5070/restriction-enzyme-digests
- **PCR**: http://www.jove.com/science-education/5056/pcr-the-polymerase-
  chain-reaction


## Procedure for Lab 1B

This dry lab is intended to be performed during your regular lab period – see the lab
schedule for details. You should plan to use the scheduled 2hr 50min lab period to
complete the steps listed in this dry lab and **also** complete and submit the
corresponding assignment during this time.

1. **Watch a short animation of genome packaging and copying**. Please access it
   at http://www.wehi.edu.au/wehi-tv/molecular-visualisations-dna

   - You may notice that the video refers to chromosomes as only being present
     at certain times during the cell cycle (i.e. when condensed). By contrast,
     tools such as the Genome Browser that you will use below tend to refer to
     'chromosomes' and 'DNA' interchangeably.

   - Note: a common question in lecture is whether the DNA helix really has a
     major and a minor groove – this can be difficult to visualize using the
     textbook figures. Look in this animation for one groove that literally spirals
     down the entire helix and is quite a bit more roomy (the 'major' groove)
     than the other – most DNA-binding proteins, such as many transcription
     factors, have a geometry that allows them to bind in the major groove to
     access DNA sequence, since the minor groove (which also spirals down the
     length of the DNA) is quite narrow. These grooves are especially evident
     during the copying part of the animation.

2. **Explore the UCSC Human Genome Browser, focusing on the intestinal alkaline phosphatase gene**. Lab 1A involved examination of a plant genome (strawberry). We'll take a look at the human genome next. Go to http://genome.ucsc.edu/ and select 'Genomes' in the top left.

- Take a moment to get oriented. First take a look at the wide variety of 'Represented Species' that you could select; a phylogenetic tree shows the calculated evolutionary relationships amongst these species based on sequence data. Hopefully there are at least some genomes in here that interest or surprise you! **You'll mention three of these organism names in your report**, along with a brief explanation of what drew your eye to those three from this long list. (Some type of personal and/or professional interest...?) Try clicking on at least one of these organisms – you'll see that it gives information about the genome sequence, including the name of the organization(s) responsible for the actual sequencing project. Near the top of the browser window is a 'Go' button that would allow you to access and browse the actual sequence information, however we are going to do this for the human genome, so proceed with the following steps.

- Go back to the top and select 'Human' from either the 'Represented Species' or the 'Popular Species' area. Over to the right, look at the 'Find Position' function at the top; the latest 'Human Assembly' should be displayed, but a drop-down menu allows researchers to choose to browse earlier drafts of the genome sequence if desired.

- Next, directly below, we need something to search for in the "Position/Search Term" box. In a later lab (Lab 3A) we'll be examining intestinal alkaline phosphatase, so let's use that. A simple keyword search returns some undesired results (try it!), so it's useful to know that the official abbreviation for this gene is **ALPI** (for alkaline phosphatase, intestinal). You should see search results in several categories (Known Genes, RefSeq Genes, Non-Human RefSeq Genes, etc.). We want a human reference sequence (RefSeq), so **look in the RefSeq Genes category** for a sequence link (gene name, chromosome number, and nucleotide position range), and click on it. (Common error: the ALPI gene is listed in other places on this page, since there are a few different types of sequence databases with information of varying levels of quality; for this exercise, specifically look for the ALPI link under the **RefSeq** category.)

- You are now in the actual Genome Browser. Take a look at the bottom part of the screen to see which data 'tracks' are shown and which are hidden – these tracks are the major advantage of the UCSC Genome Browser over the NCBI Gene database (https://www.ncbi.nlm.nih.gov/gene – also interesting!), and these other tracks are worth exploring! For most purposes, users tend to keep most tracks hidden most of the time, although of course this means that an enormous amount of genome annotation is then being hidden, simply to prevent the user from being overwhelmed with detail! (Conversely, some types of users generate many tracks of their own that they wish to display, e.g. experimental data showing for the first time the genomic locations of all the binding sites for transcription factors U, V, and W after stimulus X, Y, or Z.) For now, you should **keep all of the tracks on 'hide' <u>except</u>:** look for the following categories and use the drop-down menus to do the following:

- **In 'Mapping and Sequencing': set 'Base Position' to 'dense'** (everything else should be hidden)
- **In 'Genes and Gene Predictions': set 'GENCODE v24' and 'RefSeq Genes' to 'pack'** (everything else should be hidden)
- **In 'Expression': set 'GTEx' to 'full'** (everything else should be hidden)
- **In 'Comparative Genomics: set 'Conservation' to 'full'** (everything else should be hidden)
- **Make sure that all tracks not mentioned above are set to 'hide'.** (Common error: part of this assignment is to make sure that you can show that you know how to **exert control** over the tracks, removing undesired 'clutter'! So be sure to hide all tracks that aren't mentioned above – don't just accept whatever defaults you see.)
- Then hit 'refresh', and go back to the top of the screen.
- **If you want further guidance regarding what can be seen in the Genome Browser view, see the end of this exercise for some screenshots with explanations (using the ACTB gene as an example) before proceeding.**

- The ALPI gene is shown twice in the genome browser (with information from the GENCODE database in the top track, and with information from the RefSeq database in the second track). The two databases have almost the same sequence information, except that according to the RefSeq database the 3' untranslated region is a little longer than it is in the GENCODE database.
  - You should be able to tell where the exons of the gene are. Not only are they the thick bars (the thin lines are the introns), but you should be able to see a pattern when looking at the conservation tracks – what do you notice about the exons vs. the introns? (A tall vertical line for conservation score indicates that the sequence in that region is highly conserved amongst the 100 vertebrate genomes under comparison.)
  - Learn more about the 'Conservation' track by scrolling down to the first of the 'Comparative Genomics' tracks, and clicking on the 'Conservation' link (right above where you selected 'full' from the drop-down menu). A lot of detail is given here about how the conservation score is determined, as well as the identities of the 100 vertebrates. **Take note of three random vertebrates that catch**

**your eye from the list of 100 – you'll mention them in your report.** Then use your browser's 'back' function to go back to the Genome Browser.

- Near the top of the genome browser, look for the following information. **What chromosome is the ALPI gene on? How long is the gene in base pairs (bp)? How many exons do there appear to be? You'll include this information in your report.** Now carefully move your cursor up to the top track, which gives the base position; try to click and drag your cursor as though you were trying to select just the region containing **ONE** of the **exons**.
    - o You should find that the view expands to zoom in only on that region – so much so that you can now see the actual DNA sequence ('top' strand) near the top. Within the GENCODE and RefSeq tracks themselves is the 1-letter amino acid code for the protein (M = methionine, A = alanine, etc.), and below in the 'Conservation' area, that same protein sequence for human is compared to that of other some other vertebrates.
    - o Unsurprisingly, you can probably see that our own human protein sequence is nearly identical to that of rhesus monkey, for example. (If we were looking at an intron, then of course we would not see any amino acid sequence here – only nucleotides.)

- Click on the RefSeq track where the ALPI gene's exon is shown (it says 'RefSeq Genes' right over it); this should take you to more information about the ALPI RefSeq.
    - o **Briefly, what is the purpose of the protein that this gene encodes?** To the best of your ability, **paraphrase** using your own words – don't quote directly – for your report.
    - o Also note: **is the ALPI gene on the + or the – strand?** Chromosomes don't inherently have a 'top' or 'bottom' strand, of course, but sequencing projects use an agreed-upon convention regarding which DNA strand will be called the '+' strand and which will be the '–' strand on any given chromosome. (Approximately half of the genes in the genome are on the '+' or 'top' strand, meaning that they would be transcribed from left to right, in the view that we see them in the Genome Browser; the other half are on the '-' or 'bottom' strand, so they would be transcribed from right to left in the Genome Browser view.)
    - o Next, in the 'Links to Sequence' area, click on 'Predicted Protein' – **what is the expected length of the ALPI protein in amino acids?**
    - o Finally, find the part of the page that says 'mRNA/Genomic Alignments' and click on the long link underneath that includes the gene size, identity, chromosome number, etc. This long page first gives the cDNA sequence only (complementary to the mRNA, and

with 5'UTR and 3'UTR in red and protein-coding sequence in blue; exon borders are in lighter shades), and then aligns all of that same cDNA sequence to the genomic sequence so that you can see all of the intronic sequence as well. It is fairly common for intronic sequence to be extremely long compared to exonic sequence, but ALPI is not a particularly good example of this. **How many exons does ALPI have?** (They are called 'blocks' here, and hopefully this confirms what you counted earlier when you were in the Genome Browser view.)

- Use the 'back' function in your browser twice to get back to the Genome Browser. Now click on the other depiction of the ALPI gene – the GENCODE one. Clearly this one should have some similar information, but scroll down to see some additional links and images.
    - One is a graph of gene expression data from various cell types – **what is notable about what cell type(s) ALPI is expressed in?** (Some genes names are no longer accurate given our current state of knowledge about their function and/or location of expression; some were named for historical reasons that we now know don't accurately describe their typical expression. What can you say about the location of expression in the case of ALPI, as compared with what we would expect, <u>given its name</u>?)
    - There is some additional interesting information contained on this page. In the 'Comments and Description Text From UniProtKB' a summary catalyzed reaction is given for this enzyme (which we will study in Lab 3A), the identities of two metal cofactors are given (which we will look at in Lab 4A), and the enzyme is described as being membrane-associated through a 'GPI anchor' (which we will see depicted much later in textbook section 12.4).
    - We can also note that the 'Comparative Toxicogenomics Database' section lists one of the chemicals that binds to this enzyme as 'nitrophenylphosphate' – this is the same molecule as 'PNPP', which will be our substrate for alkaline phosphatase in our enzyme kinetics experiments in Lab 3A.
    - Finally, if you scroll down further to see the 'Predicted Comparative 3D Structures' you can see that this enzyme contains a lot of alpha-helices; we'll get a better look in Labs 4A and 5B.

- Use the 'back' function in your browser to get back to the Genome Browser again. This time, look at the track that is under the RefSeq one. It is a bar graph of gene expression data – the same one that you saw earlier when you were looking at the GENCODE track. There is no particular meaning to the locations where specific bars are found along the gene – they are just spread out along the gene that you're viewing in order to allow you to view the graph as best you can. Hover your cursor over a colored bar; it will tell you which tissue you are currently looking at. Click on it, and here there is a lot more information about where the cell-type-specific gene expression data were derived from (number, gender, and age of the tissue donors, etc.). Use the 'back' function again to return to the Genome Browser.

- Now we need to view one additional track. In the top section ('Mapping and Sequencing') find the 'Restr Enzymes' track and click on the track name. In the new screen, type 'ecori' into the 'Filter display by enzymes' box, and set the display mode to 'pack' before selecting 'submit'. (We're looking at EcoRI specifically, since the DNA Isolation and Analysis assignment involves a discussion of this enzyme.)
  - Use the zoom out '10x' button in the top right corner (you may need to do this twice) until you can see the entire ALPI gene. **Do there appear to be any EcoRI cut sites in the ALPI gene? If so, how many?** If you're not sure, zoom out another 10x. Once you start to see neighboring genes, you should definitely be seeing some EcoRI cut sites, and be able to tell whether any are in the ALPI gene itself.

- Now you are going to zoom in on a section of the ALPI gene that you would like to provide to a PCR primer-selection program. It's fine to use the entire ALPI gene, or you can zoom in on a section of (very) approximately 500bp or so (such as perhaps two or three exons from somewhere within the gene).
  - Once you can see (next to the text box at the top of the screen) that you are zoomed in on approximately 500bp or so, then you can get the DNA sequence corresponding to this displayed region. You can do this simply by looking for a menu at the top of the screen that says 'View' and then selecting the 'DNA' option. Then click on the 'Get DNA' button and the sequence will be shown, with the chromosomal location on the first line. Keep this for use in the next step.

3. **Design PCR primers capable of amplifying part of the ALPI gene**. The overall basis for PCR (polymerase chain reaction) is described near the end of section 5.1 of our textbook, and/or you can view the video explanation that is found by following the link in the 'Preparation for Lab 1B' section. In short, to allow amplification, the double-stranded DNA template is heated to separate it into single strands in order to allow the primers to hybridize, and then DNA polymerase makes new copies by template-directed synthesis. In the lecture portion of the course we will examine the active site of DNA polymerase in detail, and importantly will note that it acts by **extending from the 3' end of a primer**, which is annealed to a template.

- One of the most popular PCR primer design tools is Primer3. We will give it ALPI DNA sequence and ask it to design primers (using the default settings), and then look back in the Genome Browser to see where those specific primers anneal (i.e. bind). Our main objective here will be to visualize those primer annealing sites in the UCSC Genome Browser as well as review primer directionality conventions. Keep your Genome Browser window open (which has the ALPI DNA sequence), and open a new window where you will access Primer3 at http://bioinfo.ut.ee/primer3/

- In Primer3, locate the appropriate (largest) text box and paste in your ALPI DNA sequence that you copy from the UCSC Genome Browser (highlight it and use your browser's copy and paste functions). As you might expect, a primer design program knows that the DNA to be amplified will be double-stranded, and so you paste in a single strand and it can figure out the rest. You can leave all defaults unchanged (but if you scroll down you can see there are many options, such as for specifying optimal primer length, %GC content of the primers, product length, etc.). Select the 'pick primers' button at the left.

- On the results screen, it should have recommended a 'left' and a 'right' primer for you at the top of the page, based on the sequence that you provided. (If it didn't, then go back to the Genome Browser and obtain a longer portion of the ALPI sequence to try.) If you scroll down to the bottom you can see how many 'left' primers it considered, how many 'right' primers it considered, how many pairs it considered, and how many pairs it considered 'ok'. It should also note that many pairs were rejected because of 'bad GC%' – can you imagine the significance of having one primer with an unusually high or low proportion of G's and/or C's, assuming that you plan to use typical reaction temperatures in your PCR reaction? The key difference between G-C and A-T pairs is shown in our textbook in Fig. 1.6 (and again in Fig. 4.12).

- Let's assume that the default primers are fine for our purposes. Importantly, note that the left and right primers were not provided with any directionality! The program assumes that you know that **all primers are given in the 5'-to-3' direction**, since, by convention, this is the direction in which we would **always** expect to orient our sequences when placing a custom oligonucleotide synthesis order (i.e. primer order).
  - Knowing that the input sequence was 5'-to-3' (since this is always how it would have been downloaded by convention) and that both of the primer sequences are 5'-to-3', take a careful look at the primer sequences and where they will align with the gene template **(keeping in mind that the template would actually be double-stranded!)**.
  - Scroll down as needed – the locations where they anneal (hybridize) are marked as with >>> and <<<< symbols (one symbol per base-pair formed). You may need to sketch this out for yourself in order to keep all the reverse complementarities straight.

o   Think about this carefully and **deduce which primer (left or right) will be the one that will physically anneal (base-pair with) the actual DNA sequence that you originally provided. How do you know?** (This is important! Primer-selection results are readily confused. Think carefully about the purpose of the PCR primers, the fact that **both** of your primers are given in the 5'-to-3' direction, and the fact that the template is double-stranded. Go to the trouble of comparing the primer sequences that you are given with the marked regions in the input DNA sequence where the primers are supposed to anneal, and see whether the primers are the **same** as the input sequence or the **reverse complement** of the input sequence!)

- Now, using your two new primer sequences, go back to the UCSC Genome Browser (use the 'back' function in your browser twice if you are still on the page where you copied your DNA sequence). At the top menu in the Genome Browser there is a 'Tools' option; within this, select 'In-Silico PCR' (where '*in silico*' means computationally). This allows us to visualize where in the genome those primers would anneal, and whether they would perhaps amplify more than one region (which would be undesired – if this happened, then you would test other primer options that were considered 'ok' by Primer3).

- Retrieve your 'left' Primer3 primer sequence and paste it in as your 'forward' primer, and retrieve your 'right' Primer3 primer sequence and paste it in as your 'reverse' primer. (Do not check the 'flip reverse primer' box.) Select 'submit'.

- Determine whether the result looks promising – is there a single match to the correct chromosome (the one that we now know contains the ALPI gene)? The primer sequences should be in capital letters and the rest of the amplicon should be in lowercase letters (again – only one strand is given, with it being understood that PCR amplifies both strands). Multiple results will be given if the primers are expected to amplify multiple sites in the genome – this can sometimes happen, particularly in genomic regions that have undergone ancestral duplication events. Click on the hyperlinked chromosomal region.

- You should now see only the genomic region that would be amplified by those particular primers. **Zoom out 10x and then 3x (so 30x total) to get a better view, then take a photo of this browser view for inclusion in your report – the photo should include your student ID held up to the lower right of the computer screen, to differentiate your work from that of your classmates.** It should show the region that your PCR primers amplify (the track is called 'your sequence from PCR search').

- Then zoom out a further 100x, to gain an appreciation for the small, targeted region that your primers are expected to amplify from within such a wealth of genetic information. (You are still only looking at a tiny fraction of this chromosome! If you have a fast computer, you can continue zooming out further to get a better idea of what this chromosome's landscape is like. It is a good way to appreciate how large the chromosome is, because you get to see more and more and more genes coming into view!)

4. **Look at a gene of your choice in the Genome Browser**. Now, in the Genome Browser, it's time to leave the ALPI gene and find a different one to look at. With over 20,000 genes in the human genome, this is a chance to make your submission really unique! You can enter anything that you choose near the top of the Genome Browser where it says 'enter position, gene symbol or search terms'. Some examples include your own name (you never know! Mine is a gene name…), a gene that you've heard of, the name of a disease that you're interested in, the name of any gene/protein in our textbook, etc. **What search term(s) did you use?** Then in the search results, click on the first RefSeq link – or any one that catches your eye. (**Don't pick beta-actin** though – it is mentioned as an example at the end of this section). If you have trouble picking a gene for some reason and none of your search terms produce any results, then skip to the end of this section right now for my suggestions.) Alternatively, a completely different strategy is to type in the name of a chromosome (e.g. Chr15) and then zoom in on a random section until you encounter a gene that catches your eye.

- Take notes on the following for your chosen gene, for inclusion in your report:
    - **What is the gene symbol and the gene name? How many bp is the gene? What chromosome is the gene on?** Click on the RefSeq track to learn more about your gene – **what, briefly, is its function? How many amino acids are in the protein?** (Of course some genes do not encode proteins – if you believe that you have encountered one, then note this.)
    - **In which tissue(s) and/or cell line(s) is your chosen gene expressed most highly?** (Or the true answer may be that it is expressed similarly in all of the tissues – it's possible! Note this if it applies to you.) Now go back to the previous Genome Browser screen.

- In the Genome Browser view, zoom out 3x, so that the gene and its surroundings can be seen well. **Take a photo of this browser view for inclusion in your report – the photo should include your student ID held up to the lower right of the computer screen, to differentiate your work from that of your classmates.**
    - (Optional: although it's not required, it would be a good exercise to try copying some of the sequence from your chosen gene and asking Primer3 to generate primers, then using the In Silico PCR tool to verify

that those primers are indeed uniquely amplifying a portion of your chosen gene!)

- **General guidance regarding the tracks that we are focusing on in the Genome Browser view**, using the beta-actin gene as an example, and zoomed out 3x using the button near the top right of the screen:

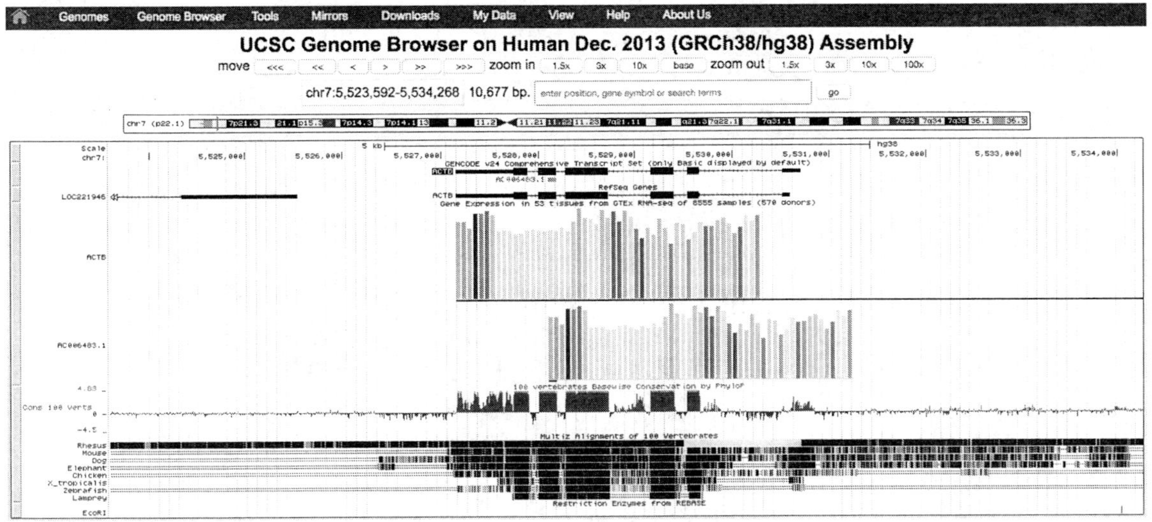

- o Searching for **ACTB** shows that beta-actin is near the 'beginning' of chromosome 7, a little past the 5,000,000[th] base-pair. (Chromosome 7 is about 159,000,000 bp in length.) The length of the ACTB gene can be given by clicking on either the GENCODE or the RefSeq track representation of the gene. In this case, the RefSeq version is slightly shorter (at 3,454 bp), since according to that database one of the exons is slightly shorter than it is in the GENCODE database. This is a 3x zoomed-out view, so it can be seen that on the right there is no gene right beside ACTB, and on the left there is a view of the end of a possible gene (LOC221946) that is found in the RefSeq but not the GENCODE database.

- o The thickest bars in the **GENCODE and RefSeq tracks** are the protein-encoding parts of the exons, while at the extreme right and left of those tracks there are medium-thickness regions that represent exonic regions that do not encode protein (i.e. the 5' and 3' untranslated regions). The thinnest portions represent the introns, and ACTB appears to have five introns. Exceptionally small and thin arrowheads on the introns point towards the 3' end of the gene, and in this case are pointing towards the left, indicating that this gene has its first exon at the far right of the screen and its sixth (last) exon at the far left of the screen. Therefore, this particular gene, by convention, is considered to be on the 'minus' strand of chromosome 7, meaning that if we were to look at the actual chromosome 7 sequence of nucleotides in any conventional genome browser starting with nucleotide 1 on the left, then we would need to take the reverse complement of that sequence in order to find sequence for the ACTB 'coding strand'. The differing lengths of the GENCODE and RefSeq versions of this gene are due to discrepancies in the two databases between where the reported 5' untranslated regions begin (i.e. where the start of Exon 1 begins).

- o The next track is the **gene expression track**. Similar data is shown both for ACTB (first graph) and for a tiny transcript known only as AC006483.1, which is found in the GENCODE database but is not considered reliable enough to appear in the RefSeq database. In other words, if this tiny gene, embedded within ACTB, truly is transcribed, it appears to be with a similar expression pattern as what is seen for ACTB. Beta-actin is a structural gene that is frequently referred to as a 'housekeeping gene', meaning that its expression is often assumed in experiments to be unaffected by treatment. A researcher might treat cells with compound X to try to block expression of a certain protein Y, and then also monitor expression of beta-actin with the aim of showing that compound X is targeted and is affecting Y only but not the expression of other proteins such as beta-actin. The gene expression track below shows that in 53 different types of cells/tissues collected from 570 different donors, ACTB expression was largely found to be expressed at similar levels, as would be expected for a structural protein that is expressed at constant levels in a wide variety of circumstances. We can see this because the 53 bars in the bar graph are of approximately equal heights; compare this pattern to what you see for ALPI! There is no significance to the left-to-right alignment of the bars along the screen – it's simply a bar graph of gene expression data for the entire ACTB gene, shown under the ACTB gene. Hovering over a bar or clicking on that track will show which bar represents which tissue/cell type, as well as numerical data from the gene expression experiment. There is some color-coding (e.g. all of the yellow bars represent various types of brain tissue), and the units for the y-axis of this bar graph are 'RPKM' – look it up if you're interested in more information on how large-scale gene expression analyses are conducted!

- o The next track is the **100 vertebrates conservation track**, and is a graph where a high value (i.e. a tall blue line at a particular nucleotide position) represents extremely high sequence conservation between 100 different vertebrate species for which sequence data is being compared. ACTB is showing a fairly typical pattern, wherein exons are highly conserved across species (as shown by the dark blue sections in the 100 vertebrates conservation track) and introns typically show lower levels of conservation. Interestingly, however, the third intron in ACTB contains a small region with unusually high conservation – a researcher might perhaps hypothesize that this intronic site is an important regulatory site for the expression of this or another gene, and the design an experiment to test this hypothesis.

- ○ Below the 100 vertebrates conservation graph are **individual alignments with the genomes of several organisms** including rhesus monkey, mouse, dog, and others. Hopefully it is unsurprising to you that we humans share substantial sequence similarity with monkey genomes (take a look at the 'Rhesus' line and see how many black lines there are), and lower (but still substantial) sequence similarity with an organism like mouse or dog. Again note that the sequence similarity is high at exons and lower in introns (with the exception being for Rhesus – with our closest relatives we have a lot of sequence similarity even in intronic regions).

- ○ The final track shows **EcoRI restriction enzyme cut sites**, which are displayed according to the directions that were provided. In this view of a 10,677 bp region of the genome, there is only one predicted EcoRI cut site, extremely far to the right. It is shown as a skinny vertical line near chromosomal position 5,534,000.

- **An example of nucleotide and amino acid sequence in the Genome Browser view**, using the beta-actin gene as an example, and zoomed in near the start of the protein-coding region in Exon 2:

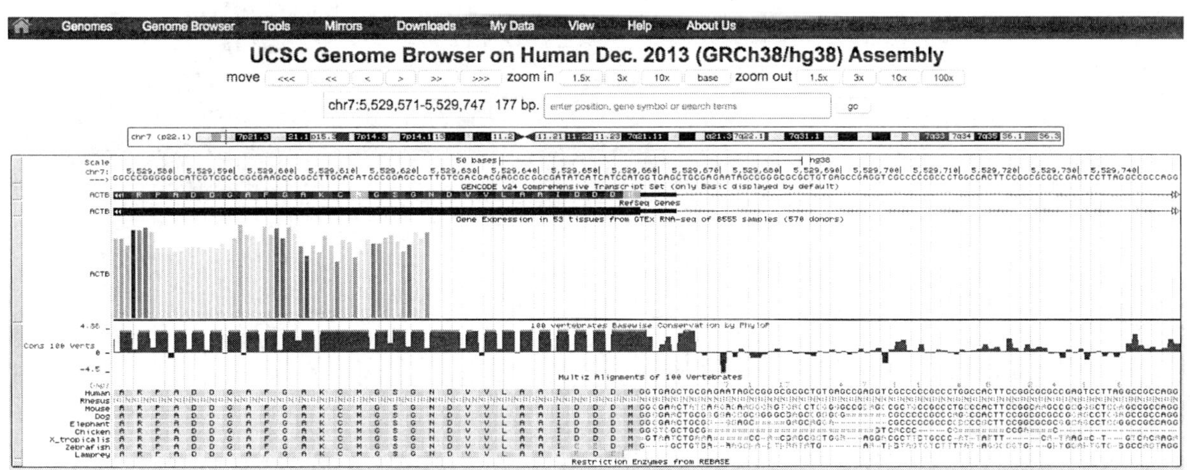

- ○ In this zoomed-in view of part of the second exon and part of the first intron, 177 bases of nucleotide sequence is visible in very small font near the top (starting on the left with 5'-GGCCC...). Only the 'top' ('plus') strand sequence of the DNA is shown; the user can figure out the sequence of the reverse complement as needed. As previously mentioned, ACTB is actually considered to be located on the 'bottom' or 'minus' strand, and so the amino acid sequence indicated below (using the 1-letter codes) would have been determined using the reverse complement of this DNA sequence. For example, the first amino acid in the beta-actin protein is M (methionine) and is shown in the approximate center of the view; the corresponding DNA sequence that is shown above is 5'-CAT-3', and the reverse complement of this is 5'-ATG-3', which corresponds to the start codon in a codon table.

- The protein sequence is highly conserved among the vertebrate sequences that are mentioned by name in this view; for example we can see that human beta-actin starts with MDDDIAALVV... (i.e. Met-Asp-Asp-Asp-Ile-Ala-Ala-Leu-Val-Val-...), and that this is also the expected start sequence for mouse, dog, elephant, chicken, *Xenopus tropicalis* (African clawed toad), and zebrafish. By contrast, the intronic sequence directly beside shows relatively little conservation at all between any of these species! (There is missing information for two species: it would require further inspection to determine the precise start locations for the Rhesus and lamprey beta-actin proteins.)

- **An example of the Primer3 primer selection tool**, using the 177 nucleotides from the previous view near the beta-actin translational start site as an example:

  - First, as described in the instructions (View -> DNA), the sequence that was visible in the current Genome Browser view (the 177 nucleotides representing part of intron 1 and part of exon 2 from the previous figure) was displayed and copied, and then pasted into the text box below in Primer3:

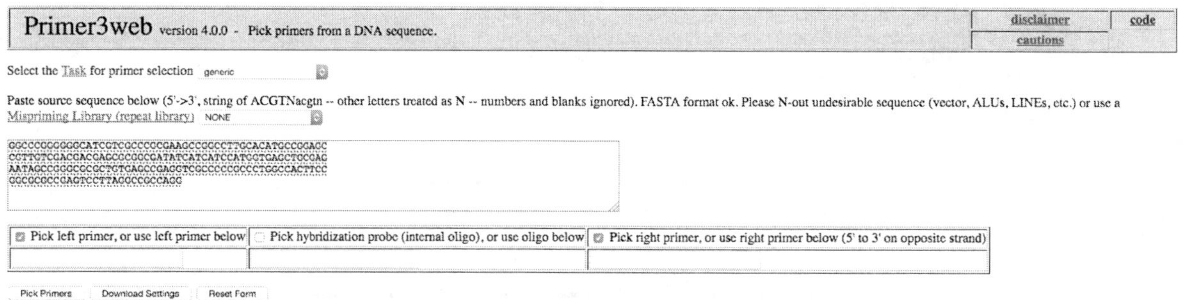



  - Second, all of the defaults were left unchanged, and the 'Pick Primers' button was selected:

## Primer3 Output

```
No mispriming library specified
Using 1-based sequence positions
OLIGO            start  len       tm      gc%  any th  3' th hairpin seq
LEFT PRIMER         68   20    58.05    50.00   12.98   0.00    0.00 GCGGCGATATCATCATCCAT
RIGHT PRIMER       176   18    59.50    66.67    1.19   0.00    0.00 CTGGCGGCCTAAGGACTC
SEQUENCE SIZE: 177
INCLUDED REGION SIZE: 177

PRODUCT SIZE: 109, PAIR ANY_TH COMPL: 0.00, PAIR 3'_TH COMPL: 0.00

    1 GGCCCGGGGGGCATCGTCGCCCGCGAAGCCGGCCTTGCACATGCCGGAGCCGTTGTCGAC


   61 GACGAGCGCGGCGATATCATCATCCATGGTGAGCTGCGAGAATAGCCGGGCGCGCTGTGA
             >>>>>>>>>>>>>>>>>>>>

  121 GCCGAGGTCGCCCCCGCCCTGGCCACTTCCGGCGCGCCGAGTCCTTAGGCCGCCAGG
                                          <<<<<<<<<<<<<<<<<<

KEYS (in order of precedence):
>>>>>> left primer
<<<<<< right primer


Statistics
          con    too     in    in    not           no    tm     tm   high  high  high       high
          sid   many    tar  excl    ok    bad     GC   too    too any_th 3'_th hair-  poly  end
          ered    Ns    get   reg   reg   GC% clamp low  high  compl compl  pin     X  stab    ok
Left      468     0      0     0     0    176    0    38   199      0     0    2     0    0     53
Right     468     0      0     0     0    334    0     7    91      0     0    0     0    0     36
Pair Stats:
considered 12059, unacceptable product size 12058, primer in pair overlaps a primer in a better pair 2150, ok 1
libprimer3 release 2.3.6
```

- o Third, **think about** how the primers are both listed 5' -> 3' (it's considered so obvious that it isn't emphasized anywhere!) and how that impacts where those primers will anneal to the **single**-stranded DNA that you pasted in. Remember, it's assumed that users will know that DNA is double-stranded with reverse complementarity and be able to account for that without prompting! There are **two essential features** of the primer pair, in order for PCR to work: one primer must anneal to each strand, and the 3' hydroxyl groups must be available to serve as attackers in the DNA polymerase mechanism of action (textbook section 4.4). Therefore, each primer's 3' end must point towards the direction of extension. Look carefully to see how both of these requirements are satisfied for the primer pair that is shown in the Primer3 output figure above.
- o Fourth, use the In Silico PCR tool in the Genome Browser to verify that these primers would only amplify one region in the genome, and to create a track that marks the 'amplicon' (i.e. the region to be amplified):

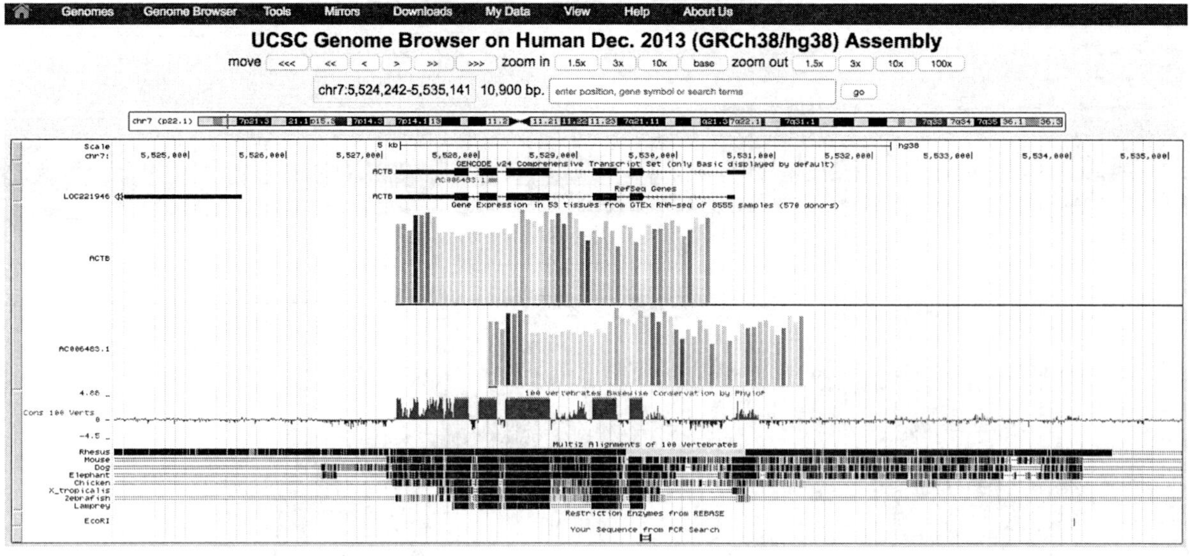

- The 'left' and 'right' primer sequences from the Primer3 output were pasted into the In Silico PCR tool as described in the instructions, and only one link was provided, indicating that only one PCR product would be expected if this primer pair were used on human genomic DNA. The amplicon is only 109 bp in length, which gives a good view of the ACTB translational start site, but not most of the rest of the gene. The view above was captured after clicking on the 'zoom out 100x' button, so that now 10,900 bp can be seen – this allows the whole gene and its surroundings to be viewed. Unsurprisingly, the PCR amplicon looks small now (newly-created bottom track), but we can still see that it is appears to be amplifying a region that spans part of the first intron and the second exon.

- The Genome Browser is an extremely useful tool, and contains an enormous number of tracks and tools that we didn't use here. Consider exploring the site's capabilities further, and perhaps also making use of the 'Help' and 'About Us' tabs to learn more.

# Lab 2A: Agarose gel of isolated DNA; protein colorimetric assay



- Take second quiz: spectrophotometry of proteins.
- Perform agarose gel electrophoresis of chromosomal DNA isolated in Lab 1A.
- Perform colorimetric assay (modified Lowry method):
    - Use pre-prepared dilutions of a standard protein sample to generate a standard curve.
    - Use the standard curve to determine the concentration of a protein sample of unknown concentration and to check for protein contamination in the DNA samples isolated in Lab 1A.

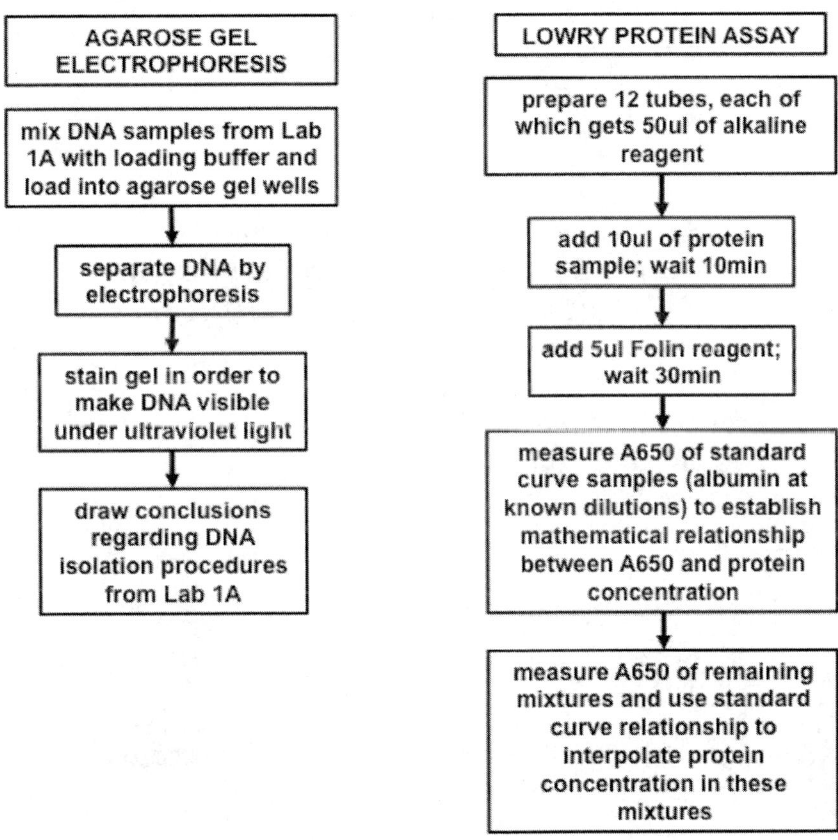

Objectives of Lab 2A

- Complete second quiz (**Quiz 2A**) on assigned reading: spectrophotometry reading from **NanoDrop Colorimetric Protein Assays Handbook (p.13, 18, and 19)**.
- Discuss challenges that arose in Lab 1A.
- Perform agarose gel electrophoresis of chromosomal DNA isolated in Lab 1A by both purification methods.
    - Run gel; take and save picture.
    - Make conclusions regarding efficacy of Lab 1A DNA isolation procedures.
- Understand basis for creating standard curve for protein determination, and associated calculations.
- Understand advantages of a colorimetric assay for protein determination.
- Note the extent to which methods for obtaining nucleic acid vs. protein concentration using the NanoDrop spectrophotometer are similar.


Preparation for Lab 2A

- Prepare for **Quiz2A** with assigned reading from the **NanoDrop Colorimetric Protein Assays Handbook (p.13, 18, and 19; provided on Blackboard)**.
- Read over the complete procedure for the lab (including both agarose gel electrophoresis and the Lowry assay).
- It is anticipated that you are already somewhat familiar with agarose gel electrophoresis from prerequisite courses. If desired, you may also review a video that discusses this technique at https://www.jove.com/video/3923/agarose-gel-electrophoresis-for-the-separation-of-dna-fragments


Procedure for Lab 2A

1. **Prepare DNA samples from Lab 1A for agarose gel electrophoresis**. Consult the 'Agarose gel electrophoresis of DNA samples' procedure that follows for further details. It is easy to get the order of samples confused – be certain to record in your lab notebook which sample is being run in which lane (Lane 1 will have the DNA ladder, and lanes 2-8 are available for running samples – be perfectly clear in your notebook which one is in which lane!). Run the gel as directed (<60min), being certain that all DNA samples are mixed with loading dye before loading. Stain the gel with GelRed and take a photo of the gel under ultraviolet light as directed. Detailed gel electrophoresis procedure follows.

2.  **Set up mixtures for the Lowry assay**. Create a standard curve using pre-prepared protein concentration solutions (20, 50, 100, 150, 200, and 500 mg/L). Then assay the unknown protein sample (which is available in undiluted, 1:10 diluted, and 1:100 diluted form) as well as the DNA from Lab 1A to check for protein contamination. Detailed Lowry procedure follows.

3.  **Clean up**. Spectrophotometers should be turned off (switch is at the back), and the upper and lower parts of the pedestal (the parts that came in contact with the protein solution) should be carefully and thoroughly wiped with a water-soaked lint-free tissue and then a dry tissue, so that no reagents dry on. Clean up your bench completely, putting the waste Lowry reagents and agarose gel reagents in the appropriate receptacles.

4.  **Look over the assignment questions**. If you have any remaining time after your lab notes are complete, look over the questions for the Protein Concentration and SDS-PAGE assignment. (Some of the questions for that assignment are on this lab, and some are on Lab 5A.)

5.  **Hand in a copy of your lab notebook pages for grading before you leave at the end of the lab period**.

    - Although you are encouraged to start the first part of the Protein Concentration and SDS-PAGE assignment early, keep in mind that the due date for this assignment will be <u>much</u> later, since it involves work that will be done in Lab 5A. It is **typical** in lab work for a long time to elapse before needing to refer back to lab notebook details – be sure that you have a clear record of your data and observations that you will be able to use when needed!

Agarose gel electrophoresis is a very popular method for the size-based separation of nucleic acids. Agarose gels are relatively inexpensive, strong, and easy to work with; the agarose is a carbohydrate polymer 'sieve' that shorter molecules travel through more readily than longer molecules do. The inherent negative charge on the phosphate backbone of the nucleic acid molecules causes them to travel to the positively charged electrode (which is called the **anode**). The gel can readily be stained with a dye such as GelRed that binds double-stranded DNA, and the DNA can then be viewed under ultraviolet light.

## Materials:

> pre-cast 1% agarose gel (8 wells; each well holds 15ul)
> TrackIt 1kb Plus DNA Ladder (Invitrogen)
> 10x loading dye containing bromophenol blue, xylene xyanol, and sucrose
> gel running buffer (i.e. TAE buffer)
> GelRed solution

## Procedure:

1. **Plan out what will go into the agarose gel.** Clarify with your TA how many strawberry DNA samples will be run on each gel, and whether any 'positive control' DNA is available. Your own sample that you isolated in Lab 1A only needs to be prepared for one lane of one gel. Then make a table in your notebook like the one below to keep track of what samples will go in each lane of the agarose gel, so that when you get the gel image, you know what is in **every** lane.

   - A positive control, in this case, would be previously-prepared genomic DNA – in other words, a sample that could be expected to yield a typical result for genomic DNA, provided that loading dye has been properly added and the gel is run and stained in the normal manner.

   - Note: if a positive control DNA sample is available, then be sure to mix it with the appropriate amount of 10x loading dye – '10x' means that it is currently in a form that is 10x as concentrated as it should be. Therefore, for example, if you are told to load approximately 5ul of positive control DNA, then 4.5ul DNA and 0.5ul of the 10x loading dye would be suitable (because the 10x dye has now been diluted 10-fold, from 0.5ul down to 5ul).

| gel lane | sample | volume of DNA | volume of 10x loading dye |
|---|---|---|---|
| 1 | DNA ladder | 5ul | none |
| 2 | strawberry DNA - precipitation method (Group 1) | 13.5ul | 1.5ul |
| 3 | strawberry DNA - spin column method (Group 1) | 13.5ul | 1.5ul |
| 4 | strawberry DNA - precipitation method (Group 2) | 13.5ul | 1.5ul |
| 5 | strawberry DNA - spin column method (Group 2) | 13.5ul | 1.5ul |
| 6 | strawberry DNA - precipitation method (Group 3) | 13.5ul | 1.5ul |
| 7 | strawberry DNA - spin column method (Group 3) | 13.5ul | 1.5ul |
| 8 | positive control DNA | see TA | see TA |

2. **Thaw the DNA sample.** Locate your strawberry DNA sample from Lab 1A, and thaw it at room temperature or with your gloved hands. Briefly ('short spin' setting) centrifuge the tube if needed to bring the liquid to the bottom of the tube so that DNA will not splash out when you open the cap.

3. **Prepare the DNA for gel loading.** Add the DNA and loading dye to a new 1.5ml tube; be sure to label it so that it does not get confused with another group's. Mix by gently flicking or vortexing, and then briefly centrifuge if needed to bring all droplets back down to the bottom of the tube.

4. **Prepare the agarose gel for loading.** Follow your TA's instructions for preparing the agarose gel for loading – wearing gloves, remove the gel from its packaging and place it in the gel apparatus, then cover the gel with running buffer.

5. **Load the gel.** Coordinate with other students to ensure that 5ul of ladder, 15ul of each strawberry DNA sample, and the positive control are carefully loaded into the appropriate lanes in the same order as planned.

6. **Run the gel.** Cover the gel apparatus with the lid, and run the gel at 100V as instructed; the power should be turned off when the dye front is approximately 3-6cm from the end of the gel (<60min). Be sure to record in your notebook how long the gel ran for! **Caution – never manipulate a gel while it's running or connected to a power source! Turn off the power and disconnect the cables before opening the chamber.**

   - Note: it's good practice to visit the gel after ~5-10min and verify that the dye front is migrating as expected, since loose connections and other equipment failures can prevent the gel from running, but can possibly be remedied when detected early!

7. **Stain the gel.** Wearing gloves, **remove the gel from the plastic holder**, and transfer it into a container of GelRed stain – this dye binds to DNA, and can be

seen when illuminated with ultraviolet light. Stain the gel in GelRed for 30min to allow the dye to diffuse into the gel and bind to the DNA, agitating it gently on a shaker. Use only containers reserved for this purpose.

8.  **Take a picture of the gel**. Wearing gloves, transfer the gel to the UV box, and take a picture as instructed. **Caution - wear safety glasses to protect eyes from damaging UV light!**

9.  **Share the gel image**. Clarify how everyone will receive an image of the gel that contains their sample, and make sure that your TA has a copy.



There are a variety of colorimetric methods for determining protein concentration, each with advantages and disadvantages. The **NanoDrop Colorimetric Protein Assays Handbook** describes several (Pierce, BCA, Bradford, and Modified Lowry), and the one that we will perform today is the Lowry method. It is also possible to get an estimate of protein concentration with an A280 measurement (see Lab 5A), though any contaminating non-protein components that absorb UV light (e.g. nucleic acids) will interfere and can introduce error in the results. By contrast, methods that rely on a chemical reaction with proteins to generate a particular colored product are expected to be more specific.

The published Lowry method is considered to be the most highly cited scientific paper in existence! The basis for this method is a two-step reaction, which takes place at room temperature:

1.  **The proteins are treated with copper in alkaline solution. This reaction is allowed to proceed for 10min.**

2.  **The proteins (primarily their aromatic side chains – phenylalanine and tyrosine) reduce the Folin reagent (phosphomolybdic-phosphotungstic acid) to produce a blue-colored product that can be detected at 650nm. This reaction is allowed to proceed for 30min.**

In a colorimetric assay, a 'standard curve' (also known as a calibration curve) is first generated, using dilutions of a protein sample of known concentration. The absorbance values are compared to the known concentrations to establish the mathematical relationship between light absorbance (at 650nm, in the case of the Lowry assay) and protein concentration. Then the A650 values for unknown protein samples can be measured, and the corresponding protein concentrations interpolated. For this to be valid, it is important that the concentrations of the protein samples lie within the range of concentrations over which the standard curve was originally generated.

## Materials:

BSA (bovine serum albumin) protein standard pre-diluted to six different
concentrations (0.020, 0.050, 0.10, 0.15, 0.20, and 0.50 mg/ml)
unknown protein solution (undiluted, 1:10 diluted, and 1:100 diluted)
strawberry DNA (precipitation method and column-purified method, from Lab
1A)
alkaline reagent
Folin reagent
1.5ml tubes

## Procedure:

1.  **Make a sample preparation table in your notebook to indicate the
    mixtures that you will be preparing.** Make a table like the following one in
    your notebook. Note that in the colorimetric assay as defined in the NanoDrop
    software, the 'blank' will be pure water, whereas tube 1 will be the 0 mg/L BSA
    sample (so the blank will be 100% water, whereas the first tube will contain
    alkaline reagent and Folin reagent). For the strawberry DNA, use the sample
    that you personally isolated, as well as some that was prepared by another
    group by the other isolation method (be sure to cite those people by name in
    your lab notebook).

| tube # | sample | volume of sample | volume of alkaline reagent | volume of Folin reagent (**wait 10min before adding!!**) |
|---|---|---|---|---|
| 1 | water | 10ul | 50ul | 5ul |
| 2 | 0.020 mg/ml BSA | 10ul | 50ul | 5ul |
| 3 | 0.050 mg/ml BSA | 10ul | 50ul | 5ul |
| 4 | 0.10 mg/ml BSA | 10ul | 50ul | 5ul |
| 5 | 0.15 mg/ml BSA | 10ul | 50ul | 5ul |
| 6 | 0.20 mg/ml BSA | 10ul | 50ul | 5ul |
| 7 | 0.50 mg/ml BSA | 10ul | 50ul | 5ul |
| 8 | unknown protein sample undiluted | 10ul | 50ul | 5ul |
| 9 | unknown protein sample 10x diluted | 10ul | 50ul | 5ul |
| 10 | unknown protein sample 100x diluted | 10ul | 50ul | 5ul |
| 11 | strawberry DNA - precipitation method | 10ul | 50ul | 5ul |
| 12 | strawberry DNA - column method | 10ul | 50ul | 5ul |

2.  **Make a results table in your notebook.** You should be prepared to record the
    data (sample concentration and the A650 value from which it was derived) for
    each of the 12 samples. In particular, note that in the NanoDrop software, the
    data for your BSA standard curve will **not** be saved – <u>record each value as you
    go</u>!

| tube # | mg/ml | A650 |
|---|---|---|
| 1 | | |
| 2 | | |
| 3 | | |
| 4 | | |
| 5 | | |
| 6 | | |
| 7 | | |
| 8 | | |
| 9 | | |
| 10 | | |
| 11 | | |
| 12 | | |

3. **Prepare the mixtures.** Number 12 tubes, then carefully add 50ul of alkaline reagent to each. Next, carefully add 10ul of the appropriate sample to each. Vortex, then **wait 10min**. Then the final step, after alkaline reagent and sample have been mixed and the 10min incubation period is over, is to add the Folin reagent, and vortex the sample again.

   - Be sure to re-freeze your DNA sample, and note where it is being stored and what is written on the label, since you will use it again much later in Lab 5A.

4. **Incubate the mixtures at room temperature for 30min.** They can incubate for longer than 30min if needed (if you are attending to your agarose gel or if the spectrophotometers are occupied). Note the time at which the incubation started and finished, and note the overall color of the mixtures. During this incubation time, you should also check on your agarose gel and/or look ahead at the assignment questions.

   - 30min is the optimal incubation time. However, if you are starting to run out of time, then decrease the incubation time (e.g. to 15-20min), since you need to allow time to read each of your samples on the spectrophotometer. Note the total incubation time in your notebook.

5. **Measure the absorbance of each sample, and record the calculated protein concentration.** Locate the Lowry colorimetric protein concentration option in the NanoDrop menu. It will first ask you to enter the concentration of all of your BSA standards (pay careful attention to the units!). Note that for the NanoDrop, 2ul is recommended for protein samples (recall that 1ul was used for DNA samples; protein samples tend to contain detergents or other reagents that interfere with water's surface tension and thus make it more difficult for a tiny volume of sample to form the necessary column of liquid between the top and

bottom of the instrument's pedestal). Also, note that the NanoDrop still recommends that pure water be used to initially blank the machine; tube 1 (water + Alkaline Reagent + Folin Reagent) is referred to as the 'zero reference', and is **not** the 'blank'.

- Be sure to record **all** data that are provided by the spectrophotometer for **each** sample. **Do this as you go along – don't wait until the end!**

- Examine the absorption spectrum for each sample as you take each measurement. Make a quick (~2min) sketch of **one** example absorption spectrum in your lab notebook, and note which sample it is (suggested: tube 7, which has the highest concentration of your BSA standard). The purpose is to give the overall shape of the absorption spectrum, and also to label the peak absorption. (Does absorption appear to be highest at 650nm or not? The default 'Lowry' program in the NanoDrop software is giving us the numerical absorption value at 650nm for our data table, so hopefully absorption is high here!)

  - It's very important to remember what the spectrophotometer is doing – showing the wavelengths of light that the sample is **absorbing**, which are the wavelengths that we can **<u>not</u>** see! Therefore, when a sample has a certain color, we expect absorption at that wavelength in the spectrum to be **minimal**, not maximal. (A different example is shown in your textbook in Fig. 19.27; the spectra are shown for two different types of chlorophyll molecules, which are the green pigments in leaves. Because we see these pigments as green, they have absorption **peaks** around 400-500nm (violet-blue) and 600-700 (orange-red) and almost **no** absorption at 500-600 (green-yellow)!)

- Recall from the quiz reading that 'unusual spectra' patterns were depicted p.19 of the NanoDrop Colorimetric Protein Assays Handbook. **If any of your spectra display any of these 'unusual' patterns, be sure to note this, so that you will be able to speculate on the likely cause** (as identified on p.19).

- Since you indicated to the NanoDrop software which samples represent your standard curve and what the protein concentration is in those samples, the software automatically generates a standard curve plot, which is a scatterplot of protein concentration vs. A650. It displays the $r^2$ value, which you should look for and **record** in your notebook – this is a statistical measure of how close your standard curve is to perfect linearity. A value of 1.0 would indicate a perfectly straight line; assuming that the BSA dilutions were prepared with a high degree of accuracy, then the $r^2$ value largely serves as a measure of your pipetting accuracy as you prepared your mixtures.

6. **Note which samples have protein concentrations that fall outside of your standard curve.** Using the standard curve data, the NanoDrop software calculates the equation of the line that relates A650 to protein concentration. It then uses this mathematical relationship to calculate a protein concentration for each of your 'unknowns' (i.e. samples that you measure after your standard curve measurements are complete). It should only do this, however, for unknowns that have A650 values that lie **within** the range of absorbance values spanned by the standard curve. Therefore, make special note of any samples that had protein concentrations that were too high to be calculated, due to the limited range of the standard curve that we had.

References for Lab 2A

This lab procedure was modified from the earlier version of the course manual in order to incorporate use of the NanoDrop instruments.

Kresge, N., Simoni, R. D., Hill, R.L. (2005). The Most Highly Cited Paper in Publishing History: Protein Determination by Oliver H. Lowry. Journal of Biological Chemistry, 280(28), e26-e28.

# Lab 2B: Introduction to Jmol and protein calculator tools


Overview of Lab 2B

- Discover what information is contained in a PDB (Protein Data Bank) file.
- Use Jmol to visualize molecular structures in 3D and perform a wide variety of manipulations.
- Gain a better visual understanding of protein composition, conformation, folding, and interior view.
- Use a web-based tool to perform calculations on a sequence (molecular weight, amino acid composition, pI, etc.).
- Explore the relationship between protein sequence and pI.


Objectives of Lab 2B

- Learn a bit about the vast amount of information that can be found at the PDB website.
    - Be aware that the PDB is freely and publicly available, serving as the single repository of information about the 3D structures of proteins, nucleic acids and complex assemblies
    - Learn how to locate and download a PDB file, and view some of the associated information that is available on the PDB website
- Learn about some of the many capabilities of Jmol, a freely-available open-source molecular visualization program.
- Explore protein secondary and tertiary structure in detail.
    - View the specific H-bonds that hold the main types of secondary structure together
    - Compare various common types of visualization (cartoon, stick, wireframe, spacefill, surface)
    - Explore various ways of coloring molecules (e.g. by element type in order to view H-bond donors/acceptors; by chemical properties in order to view nonrandom localization of R-groups)
    - View the R-group chemistry on the exterior and interior of a globular prpotein
- Go to the ExPASy (Expert Protein Analysis System) website and use the ProtParam tool to view protein parameters after providing a protein sequence (or identifier), including:
    - Length
    - Molecular weight and formula
    - Residue composition
    - pI
    - Extinction coefficient

Jmol is a freely-available program that will be required for some of these exercises. There are much more complex programs available, however Jmol is a small, relatively simple program that will meet our needs, and allow us to follow the instructions as we complete selected exercises that have been modified from the manual "***Studying Protein and Nucleic Acid Structure with Jmol. For use with Berg, Tymoczko, Gatto, and Stryer, Biochemistry, 8th Edition.* By Jeffrey A. Cohlberg. Revised May 2015**".

Before moving on, be sure that you have downloaded Jmol from http://jmol.sourceforge.net/ using the instructions provided there (note that you must have Java installed on your computer in order to run Jmol). You should also have completed the BIOL 3611 lecture assignment "**INTRODUCTION TO JMOL – DNA**".


Procedure for Lab 2B

This dry lab is intended to be performed during your regular lab period – see the lab schedule for details. You should plan to use the scheduled 2hr 50min lab period to complete the steps listed in this dry lab and **also** complete and submit the corresponding assignment during this time.

1. Using Jmol, complete **EXERCISE 1: SOME BASICS OF SECONDARY AND TERTIARY STRUCTURE**, which was modified from the corresponding exercise in "Studying Protein and Nucleic Acid Structure with Jmol". The modified instructions appear here – note that you will have only one Jmol file open at a time. **Finish up your responses for one Jmol file before opening a new one, so that you don't lose any of your work!**

   - To complete **EXERCISE 1: SOME BASICS OF SECONDARY AND TERTIARY STRUCTURE**, please proceed with the following steps:

     i. Go to http://www.rcsb.org/pdb/ to access the Protein Data Bank. This resource contains a wealth of information about proteins and other macromolecules, including the detailed structural information for many thousands of molecules.

     ii. Go to the search box in the top right and type in **1AL1** – this should take you to the record for a small alpha-helical peptide. There are tabs at the top (Structure Summary, 3D View, Annotations, Sequence, and others) that give further information regarding what is known about this database entry, as well as relevant links to other database entries. The Structure Summary page includes a link to the journal article in which this structure originally appeared, the complete sequence of this 13-amino acid peptide (1-letter amino acid code), and a 'Small Molecule' section that indicates that the peptide was crystalized along with a sulfate ion and an acetate. The 'Experimental Data' section indicates that the structure was solved by x-ray crystallography, and gives the resolution in Angstroms as well as additional experimental information.

iii. In the top right, click **Download Files -> PDB Format (gz) -> Save File** to download the 3D structure file. It is a compressed file, but you don't need to expand it before opening it in Jmol – you just need to note where on your computer it is saved.

iv. Open Jmol (using the .Jar file or Executable file, as always), and if the console isn't already showing then in the top menu go to **File -> Console** to display the Jmol Script Console.

v. Type **set pdbAddHydrogens** – this will ask Jmol to show hydrogens in the structure, even though they could not be resolved in the original structure data due to technical limitations. You should receive the response 'pdbAddHydrogens = true'.

vi. In the top menu go to **File -> Open** and locate the 1AL1.pdb.gz file that you just downloaded. It should open right away to show a right-handed alpha-helix, a sulfate ion, and an acetate. (It's important that the image appear in the Jmol window, not a web browser window. If you are accidentally viewing this file in a web browser, go back and follow the instructions for opening it in Jmol, so that you will be able to follow the directions as we move forward.)

vii. You can rotate the structure in the X and Y directions (left-right and up-down) by left-clicking on the molecule and dragging on it. You can move it in the Z direction (toward you and away from you) by holding down the 'shift' key and moving your cursor up or down. You can translate the entire molecule (move it around) – try holding down the 'shift' key and left-clicking twice – on the second click keep your finger held down and move the cursor around.

viii. The pop-up menu has many useful commands – get to it with a right-click (PC) or two-finger click (Mac). Try **Style -> Scheme -> CPK Spacefill** and then **Style -> Scheme -> Ball and Stick** – both are great for showing the elemental composition of the alpha-helix but make it more difficult to see its orderly helical structure. Notice that the Style menu has many additional options (e.g. the thickness of the bonds can be changed; by default in the Ball and Stick view the bonds are 0.15 Angstroms).

ix. Now from the pop-up menu try **Spin -> On** and then click on the molecule. Then stop it with **Spin -> Off**.

x. Next from the pop-up menu try **Surfaces -> Dot Surface** for an interesting effect. Then **Surfaces -> Off** then **Surfaces -> van der Waals Surface** then **Surfaces -> Make Opaque** for one of the most realistic looks at the space that this alpha-helix would actually be likely to occupy if it were found in a protein. Then choose **Surfaces -> Off**.

xi. Now make sure you are back at **Style -> Scheme -> Ball and Stick** and then we'll switch to using the console for our next set of commands.

xii.    In the console, type **ribbon** – this gives the path of the helix but still shows the ball-and-stick view at the same time. Now type **ribbon only** and then make it narrower with **ribbon 200.**

xiii.    Now type **ribbon off; backbone 100** – rather than a smooth ribbon, this view gives straight lines connecting each of the alpha-carbons. If you rotate the structure so that you are looking down the axis of the helix, you should be able to see that there are between 3-4 amino acid residues per turn of the helix, based on the path that the backbone is taking.

xiv.    Now use the pop-up menu to display **Style -> Scheme -> Ball and Stick** again. If you're still looking down the axis of the helix, you should be able to see that the side chains extend outwards from the helix.

xv.    We can color-code these side chains by their chemistries. In the console, type **select hydrophobic; color green** and then **select positive; color yellow** and then **select negative; color orange** – we can see that this is an 'amphipathic' helix, with one side having hydrophobic side chains and the other side having charged side chains. (In the extended <u>linear</u> peptide sequence, of course, we wouldn't expect all of the hydrophobic and all of the charged residues to each be clustered together – see textbook Figs. 2.24 and 2.25 if needed.)

xvi.    To emphasize the coloring further type **select protein; spacefill**

xvii.    The sulfate and the acetate are a bit distracting – type **delete so4** and **delete ace** to remove these.

xviii.    Now let's explore the chemistry of these side chains further. (You can also click on any side chain and the 3-letter code of the corresponding amino acid will show up in the console.) Type **select hydrophobic; spacefill off; wireframe 100; color cpk** – it should be easy to see that these side chains are all hydrocarbons, based on their coloring. Their lack of nitrogen or oxygen (except in the backbone) indicates that they interact poorly with water and other polar or charged substances.

xix.    Select the **undo** button on the console.

xx.    Type **select positive or negative; spacefill off; wireframe 100; color cpk** – this view should look quite different, due to the presence of acidic and basic residues that contain interesting functional groups (most readily revealed here by their oxygen and nitrogen content; however as a side note, if you truly tried to compare the side chain structures to the structures in the textbook, you would find that this crystal structure includes some missing details, since two of the lysines could not be clearly identified in this experiment).

xxi.    Now it's time to look at a slightly larger structure, as a simple example of tertiary structure. Type **set pdbAddHydrogens**

xxii.    Go to **File -> Get PDB** and type in **1PGB** (note that you could also have downloaded this file from the Protein Data Bank website as we did earlier for the alpha-helix).

xxiii.    In the console, type **show info** – this gives us some information about the structure, including the fact that it consists of 56 residues of 'Protein G' (the PDB website record would certainly give us more information about this immunoglobulin binding protein), and contains one alpha-helix and four beta-strands. Are there any sulfurs in this protein? If so, they can be easily identified by their yellow color.

xxiv.    Type **delete water**

xxv.    Type **cartoon only** – you should now clearly be able to see the alpha-helix and the beta-sheet. Rotate the structure around to get a good look.

xxvi.    Type **color group** – now the N-terminus is in blue and the C-terminus is in red, with the colors gradually changing in between. This color scheme can be very helpful for following the backbone when its path is complicated.

xxvii.    Each arrowhead points to the C-terminal end of a beta-strand. You should be able to see that this structure contains an example of a 'mixed' beta-sheet (textbook Fig. 2.33).

xxviii.    Type **backbone only; backbone 100**

xxix.    Type **color cpk**

xxx.    To see the H-bonds, type **hbonds calculate** – then, because they appear to be floating in space (since the amino hydrogens and carbonyl oxygens from the backbone are not shown), type **set hbonds backbone**. Make them thicker with **hbonds 40**. As you probably expected, red is the oxygen end of the H-bond and white is the hydrogen end. It probably calculated a few bonds that you didn't expect based on the secondary structure coverage in textbook section 2.3 (especially in the beta-sheet – this will be fixed later with the 'set bonds sidechain' command), however the alpha-helix H-bonds should already show a very striking pattern that is completely distinct from what is seen in the beta-sheet.

xxxi.    Type **select all; wireframe only; wireframe 80** to display all of the atoms – then **set hbonds sidechain** to more accurately connect the H-bonds between the appropriate atoms. They are difficult to see, so we'll look at the beta-sheet alone, then the alpha-helix alone.

xxxii.    Type **select sheet; color cpk; restrict sheet** then hide the alpha-helical H-bonds with **select not sheet; hbonds off**

xxxiii.    This view shows the extensive and regular H-bonding between strands in the beta-sheet. Remind yourself where the strands are with **select sheet; cartoon** – remember, the side chains extend above and below the sheet (textbook Figs. 2.30-33)! They are somewhat obscuring the view of the H-bonds, so type **hide sidechain** for a beautiful view of the H-bond stabilization in this sheet.

xxxiv.    This is a great place for a photo! Change the background to any other color (see http://jmol.sourceforge.net/jscolors/#JavaScript%20colors or the lab manual appendix) using the command **background X** to change it to color X. Then take a photo of your screen that shows three things: your beta-sheet on your chosen background, your console, and your student ID in the lower right. This will be part of your assignment submission.

xxxv.    Finally, take a look at H-bonding in the alpha-helix. Type **select all; wireframe only; wireframe 80** and then **select not sheet; hbonds on;**

**hbonds 40** – this is a fantastic view that really highlights the H-bonds in both the beta-sheet and the alpha-helix. Other types of noncovalent interactions stabilize this structure as well, but the extensive H-bond stabilization is very nicely on display in this view.

xxxvi. Now focus on the helix: type **restrict helix** and then **select not helix; hbonds off**

xxxvii. If you click individually on two residues that are connected by H-bonds the names and locations of those residues will appear in the console, and you can verify that H-bonds occur between residues that are four residues apart (as shown in textbook Fig. 2.25). As an example, Thr 25 can be shown to be H-bonded to Val 29.

xxxviii. Finally, type **select all; hbonds 40; wireframe off** – isn't it interesting that you can tell where the alpha-helix and the four beta-strands are with absolutely <u>no</u> atoms shown...??

2. Using Jmol, complete <u>**EXERCISE 2: LYSOZYME, A GLOBULAR PROTEIN WITH DISULFIDE BONDS**</u>, which was modified from the corresponding exercise in "Studying Protein and Nucleic Acid Structure with Jmol". The modified instructions appear here – note that you will have only one Jmol file open at a time. **Finish up your responses for one Jmol file before opening a new one, so that you don't lose any of your work!**

- To complete <u>**EXERCISE 2: LYSOZYME, A GLOBULAR PROTEIN WITH DISULFIDE BONDS**</u>, please proceed with the following steps:

    i. In the Jmol console, type **set pdbAddHydrogens**

    ii. Go to **File -> Get PDB** and type in **2LYZ** (note that you could also have downloaded this file from the Protein Data Bank website as we did earlier for the alpha-helix). You can see from the console information that it is the structure of the hydrolase known as lysozyme from hen egg-white.

    iii. Type **delete water**

    iv. You should be able to easily see the yellow disulfide bonds.

    v. Type **spacefill** – then rotate the protein so that you can see the deep cleft where this enzyme binds its substrate (a hexasaccharide). This enzyme is involved in cleaving bacterial cell wall polysaccharides.

    vi. To see the disulfide bonds more clearly, type **backbone only; backbone 80** then **ssbonds on; ssbonds 60**

    vii. Disulfide bonds occur between side chain (not backbone) atoms; to prevent them from appearing to float, type **set ssbonds backbone** – it should now be very easy to see how many disulfide bonds there are, and where they are within the protein. You can click on either end of a disulfide bond to show in the console which residues are participating in the disulfide linkage; one example is Cys 76 with Cys 94. (On the PDB website, these disulfide linkages are depicted graphically on the 'Sequence' tab for this protein.)

viii. Now change the display to **cartoon only; color structure** – this view shows that lysozyme contains three alpha-helices (magenta), five beta-strands (yellow), and one unusual $3_{10}$ helix (purple; a different geometry than an alpha-helix, and not included in our textbook coverage).

ix. Only one type of secondary structure really lines the substrate-binding cleft – can you see which type by rotating the protein around? If not, type **spacefill** and then see which color (besides white) lines the cleft. Then click '**undo**' (or type **undo**) in the console.

x. Next we'll look at the chemistries of the side chains in lysozyme. First type **wireframe only; wireframe 100; color cpk** – from this view the only thing that we can easily see is that some side chains at the surface of the protein appear to have functional groups that would be consistent with a protein that is water-soluble (since in CPK coloring, the red is oxygen and the blue is nitrogen).

xi. Type **select hydrophobic; color green; select polar; color yellow** – do the polar R-groups tend to be on the exterior?

xii. Type **spacefill** which, since only polar residues were selected, should help emphasize any differences in spatial distribution between polar and nonpolar residues.

xiii. Now we'll make the reverse image. Type **select all; spacefill; select polar; wireframe only; wireframe 100** – what do you think of this distribution of polar vs. hydrophobic residues? Is there evidence that this protein largely contains a hydrophobic core? (The aggregation of hydrophobic R-groups is involved in the folding of 'globular' ('roundish') proteins, as discussed in textbook section 2.4.)

xiv. Next let's subdivide the 'polar' residues by whether or not they are acidic, basic, or 'polar uncharged'. Type **select positive; color red; select negative; color blue** – the likelihood is that all or most of these charged residues are in contact with the polar surroundings (since this is a soluble protein), rather than being located in the interior – but how can we verify this?

xv. We can look at a view that shows a slice through the interior, similar to textbook Fig. 2.44b, in order to emphasize two important features: both that the interior contains many hydrophobic residues and that the atoms are neatly packed (i.e. that there are many favorable van der Waals interactions that help to stabilize the folded protein). Type **select all; spacefill** then **slab on; slab 50** – this shows a slice 50% of the way through. (You can see what happens if you type 'slab 25' or 'slab 80' instead.) Be sure to use your cursor to rotate the protein (or type 'spin on') for an interesting effect! If you want to change the slab all the way from 100 to 0 and back again, hold down 'control' and 'shift' while you move your cursor up and down.

xxxix. Use your favorite background color from before or choose a new one. Then take a photo of your screen that shows three things: a slab view of the interior of lysozyme, your console, and your student ID in the lower right. This will be part of your assignment submission.

3. **Learn about a web-based tool that provides quick information about protein composition: ExPASy's ProtParam tool.**

- First we need to note that the PDB webpage that allowed the download of the 2LYZ file that was used for Exercise 2 also indicates that more information about this protein can be accessed by clicking on the protein "Accession" (AC) identifier **P00698** (those are zeroes, not letters!). Certainly that information includes the entire protein sequence – and we will use this identifier to learn more about the composition of lysozyme.

- Go to http://web.expasy.org/protparam/ and enter **P00698** in the top text box. Click on 'compute parameters'. At the bottom of the next page, click 'Submit' to confirm that you want to consider the entire protein rather than a subset.

- The result confirms the number of amino acids, provides the calculated molecular weight (in Daltons), and calculates the estimated pI (recall, pI is the expected pH at which the molecule will have a net charge of zero). **What is lysozyme's molecular weight in kilodaltons (kD – not Daltons)? Which one amino acid does lysozyme contain the most of (or does it have approximately equal representation of all of them)? What is the expected pI of this protein?**

- Lysozyme's pI is somewhat close to 7 because it has a fairly equal proportion of acidic and basic residues. If you want to experiment with extremes, see what pI you get if you go back to the input page and enter 'EEEEEEEEEE' (i.e. ten glutamates in a row, using the 1-letter code). Then try 'RRRRRRRRRR (i.e. ten arginines in a row). And then 'ERERERERER'.

- DNA wraps around histone proteins as part of the mechanism that keeps it wound in an orderly and compact way. Histones have very extreme pI's that are related to their function in attracting DNA and keeping it tightly wound. Knowing what we know about DNA and amino acid chemistry, would we expect a low or a high pI for a protein that is electrostatically attracted to DNA? Check your answer by entering **P68431** (which is human Histone 3) into the ProtParam tool. Look at the pI and then look specifically for the two acidic amino acids and the two highly basic ones – **do you notice any pattern that can help you to explain both this protein's pI and how this protein is able to perform its function well?** (You may notice that the page separately lists the number of negatively and positively charged residues underneath where it shows the individual amino acids.)

- Incidentally, note that the results page also includes a calculated extinction coefficient – this could be used in spectrophotometry, rather than using the default extinction coefficient, which is often the value for bovine serum albumin (BSA). (This is relevant if using the A280 absorption method to calculate protein concentration, which we will do in Lab 5A.)

# Lab 3A: Kinetic study of alkaline phosphatase



- Take third quiz: alkaline phosphatase assay.
- Use hands-on activities to learn about some challenging enzyme kinetics concepts that can be perplexing when encountered lecture setting alone.
- Create an accurate standard curve.
- Observe rapid enzyme-dependent dephosphorylation of *p*-nitrophenol phosphate to form a colored product, through spectrophotometric monitoring of visible wavelengths.
- Contribute to the creation of an alkaline phosphatase kinetics dataset that the class will use in Lab 4A for kinetics calculations.
- Study a mammalian enzyme of medical interest.

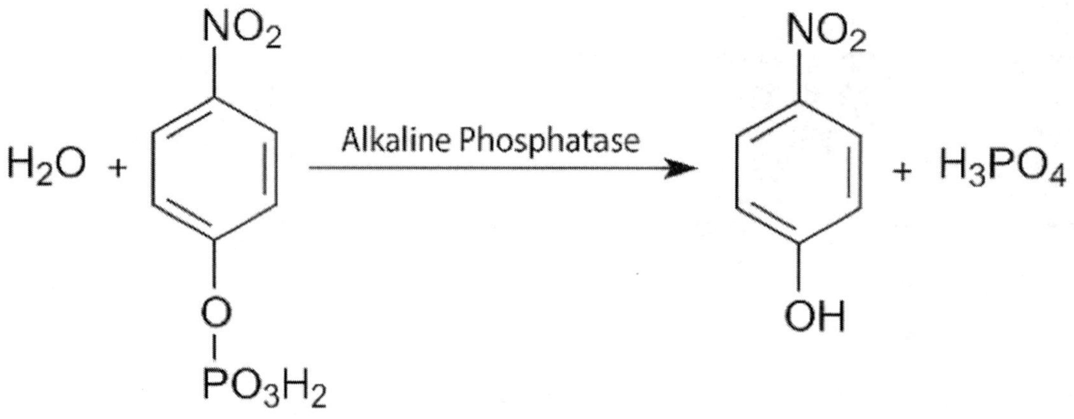

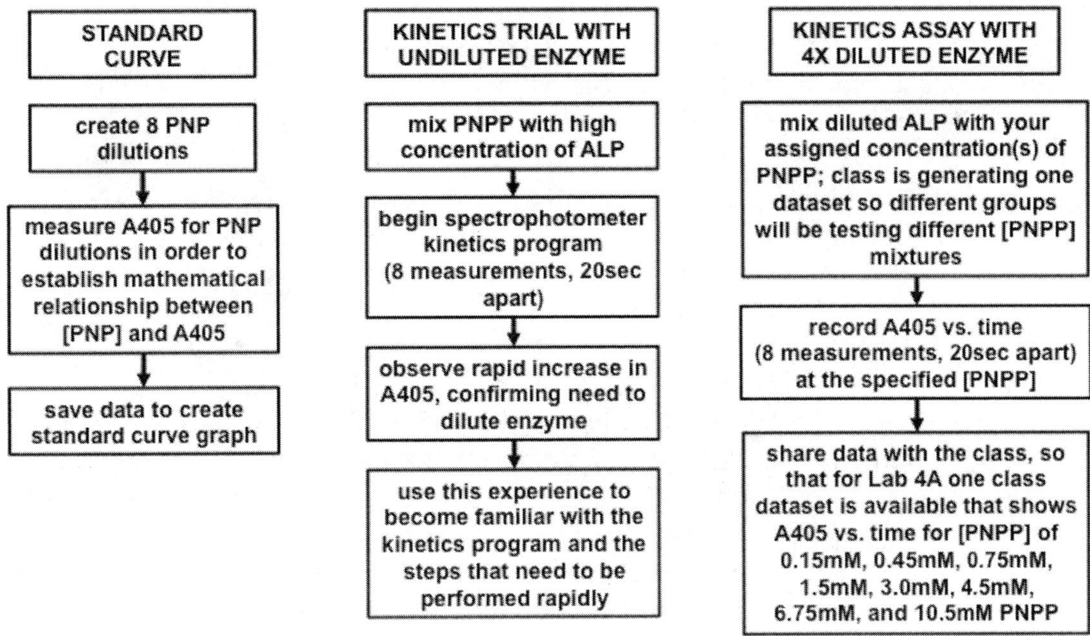


## Objectives of Lab 3A

- Complete third quiz (**Quiz 3A**) on assigned reading indicated in the 'Preparation for Lab 3A' section.
- Explore the kinetics of mammalian intestinal alkaline phosphatase, an enzyme that is widely expressed in many organisms, is readily available, is relatively stable at room temperature, is known to obey Michaelis-Menten kinetics (Section 8.4 of our textbook), is known to be inhibited by multiple types of reversible inhibitors (Section 8.5 of our textbook), and that was the focus of Lab 1B (exploration of the ALPI gene using the Genome Browser).
    - Prepare a standard curve of A405 vs. [PNP] in order to be able to determine the mathematical relationship between A405 readings and *p*-nitrophenol concentration in Lab 4A (since PNP is the product of the reaction under consideration today).
    - Prepare an initial kinetics reaction with high enzyme concentration, in order to develop familiarity with the assay and the spectrophotometer cuvette function, and to observe how overly high enzyme concentration can impede one's ability to capture the desired **linear** increase in absorbance over the specified time period.
    - Use a diluted enzyme concentration [E] to examine enzyme kinetics over a range of substrate concentration [S] values – different class members will be assigned different substrate concentrations.

- o Generate a dataset that will be used in Lab 4A to construct a Michaelis-Menten plot (**[S] vs. $V_o$**) and a Lineweaver-Burk plot (**1/[S] vs. 1/$V_o$**), to allow an estimation of $V_{max}$ and $K_m$, two parameters that report on separate attributes of an enzyme's effectiveness.
- Introduce/review key enzymatic/kinetic parameters that are challenging to learn about in the lecture setting alone. The purpose of collecting the class kinetics dataset will be to become familiar with the following parameters in Lab 4A:
  - o **Enzyme concentration [E]**: concentration in M (or mM, uM, nM, etc.) of the protein that is catalyzing the reaction.
  - o **Substrate concentration [S]**: concentration in M (or mM, uM, nM, etc.) of the molecule that the enzyme will convert into product.
  - o **Reaction velocity $V_o$**: rate at which product is formed in M (or mM, uM, nM, etc.) of product per unit of time. Is much less than $V_{max}$ when substrate concentration is low; at low [S] most copies of the enzyme can be expected to have an empty active site at any given time.
  - o **Maximal velocity $V_{max}$**: the theoretical maximal velocity that can be achieved without changing any reaction parameters (such as adding more enzyme).
  - o **$K_m$ (the Michaelis-Menten constant)**: the [S] required to achieve a velocity that is half of $V_{max}$ – and so it must have the same units as [S] does. $K_m$, like $k_{cat}$, is an inherent and widely reported property. $K_m$ is an indicator of how readily the enzyme binds to the substrate – a low value indicates that even at low [S] the enzyme can still readily achieve ½ $V_{max}$, whereas a high value indicates that it is hard for the enzyme to achieve ½ $V_{max}$ until the system is becoming flooded with substrate.
  - o **$k_{cat}$ (the turnover rate)**: the calculated number of reactions that one copy of the enzyme can possibly catalyze per unit of time. It has official units of 'per unit of time' (e.g. $s^{-1}$), and unofficially you could think of this as 'reactions per second'. As you might expect, this is calculated as $V_{max}$ divided by [E], which is how rapidly product is forming in your reaction, divided by how much enzyme there is. (In other words, it is $V_{max}$ on a per-enzyme basis.) Whereas $V_{max}$ is specific for your experimental setup, $k_{cat}$ is an inherent, widely reported property of an enzyme.
- Develop an appreciation for alkaline phosphatase's high velocity, in preparation for Lab 4A's in-depth look at the mechanism and key R-groups in the active site of this enzyme.

- Prepare for **Quiz3A** by carefully reading the boxed 'Preparation for Quiz3A' reading that follows.
- Read over the complete procedure for the lab.

---

**Preparation for Quiz3A**

**Is this enzyme found in nature?**

Intestinal alkaline phosphatase (IAP) is encoded by the ALPI gene, which we studied in Lab 1B. This is one of many phosphatases that are widely expressed across a wide variety of life forms; some are highly specific for a certain substrate, but intestinal alkaline phosphatase (ALPI) has fairly broad specificity – meaning that it can remove a phosphate group from many different substances. The 'alkaline' in its name refers to its increased activity at high pH.

**Will we be studying the reaction that IAP is catalyzing in our bodies right now?**

No – for convenience we will study a reaction that is specifically chosen because it produces an easy-to-measure colored product (rather than one that is physiologically relevant). We will monitor the dephosphorylation of colorless p-nitrophenyl phosphate (PNPP) to form yellow p-nitrophenol (PNP).

$H_2O$ + [p-nitrophenyl phosphate structure: benzene ring with $NO_2$ at top and $O{-}PO_3H_2$ at bottom] $\xrightarrow{\text{Alkaline Phosphatase}}$ [p-nitrophenol structure: benzene ring with $NO_2$ at top and $OH$ at bottom] + $H_3PO_4$

**Is there a medical reason to study this enzyme?**

IAP is being studied as a potential treatment for a variety of conditions, including inflammatory bowel diseases. IAP is involved in fatty acid absorption in the gut and maintenance of normal gut microbiota; knockout mice unable to express IAP are prone to altered metabolism and obesity.

**What is the main outcome of a kinetics experiment?**

Enzyme kinetics involves the study of the rate at which product is formed. Enzymes obey the laws of thermodynamics and therefore do not change the $\Delta G$ of a reaction, however they are capable of speeding up the rate of a reaction perhaps billions or trillions of times compared with the uncatalyzed rate (it depends on the enzyme in question). We will frequently refer to reaction velocity ($V_o$), which refers to concentration of product formed per unit of time. We will measure formation of the yellow product (PNP) via its absorbance at 405nm.

**Are there any particularly hazardous reagents in this experiment?**

AVOID contact with PNP, the yellow product. Wear gloves and eye protection.

---

## Procedure for Lab 3A

1. **Prepare the mixtures for the standard curve.** Prepare a table like the one that follows, with an empty column for the A405 absorbance data. Label eight 1.5ml tubes and carefully measure out the appropriate volume of alkaline buffer into each and then PNP stock into each.

   - **PNP is toxic – wear gloves and eye protection while working with the PNP stock solution.**

- Note: the final [PNP] in each mixture is already calculated for you below; you will need this in order to make your standard curve graph. Final [PNP] can be calculated by noting the original PNP concentration (200uM) and dividing it by the dilution factor. For example, when 25ul of PNP stock is being diluted into a total volume of 25ul + 975ul = 1000ul, then the PNP is being diluted 1000ul/25ul = 40-fold. Therefore, in tube 1 the 200uM PNP stock has been diluted 40-fold to 200uM/40=5uM.

| tube | volume of 200uM PNP | volume of alkaline buffer | final [PNP] | A405 |
|------|------|------|------|------|
| 1 | 25ul | 975ul | 5uM | |
| 2 | 50ul | 950ul | 10uM | |
| 3 | 75ul | 925ul | 15uM | |
| 4 | 100ul | 900ul | 20uM | |
| 5 | 125ul | 875ul | 25uM | |
| 6 | 150ul | 850ul | 30uM | |
| 7 | 175ul | 825ul | 35uM | |
| 8 | 200ul | 800ul | 40uM | |

2. **Measure the A405 of the standard curve.** Measure the absorbance using the 'custom' protein measurement option, entering 405nm as the wavelength that you wish to measure.

- Blank the instrument with alkaline buffer. Use the pedestal area as you did in previous labs, and use 2ul of each mixture for your measurements; it is not necessary to use the spectrophotometer's cuvette function until performing the actual kinetics measurements.

- Make a quick (~2min) sketch of **one** example absorption spectrum in your lab notebook, and note which sample it is (suggested: tube 8, which has the highest concentration of PNP). The purpose is to give the overall shape of the absorption spectrum, and also to label the peak absorption. (Does absorption appear to be highest at 405nm or not? The 'Alkaline Phosphatase kinetics' program in the NanoDrop software is going to give us the numerical absorption value at 405nm for our data table to indicate enzymatic production of PNP, so hopefully PNP absorption is high here!)

3. **Perform an initial trial of the kinetics experiment with high enzyme concentration.** In a 1.5ml tube, mix together 600ul of alkaline buffer and 375ul of substrate (PNPP; stock concentration 30mM). Then go to the NanoDrop, under the Kinetics tab select the Alkaline Phosphatase program that will take nine measurements 20sec apart, and blank the instrument with a cuvette containing 1ml of alkaline buffer. Then finally, add 25ul enzyme to your reaction mixture, and transfer it using a 1000ul pipettor to a cuvette (trying to be quick and yet trying to avoid introducing bubbles). Have a TA show you how the cuvette goes into the cuvette holder of the spectrophotometer – there is a

correct and an incorrect orientation for the cuvettes. Then begin the measurements, which should start after a short 3sec lag, after you tell the software to begin measuring. Record the A405 values in your notebook.

| time (sec) | A405 |
| --- | --- |
| 3 | |
| 23 | |
| 43 | |
| 63 | |
| 83 | |
| 103 | |
| 123 | |
| 143 | |
| 163 | |

- Note: this step serves two purposes. First, it allows you to become familiar with the overall setup for the enzyme kinetics experiment (including the use of the NanoDrop cuvette function, the kinetics option in the software, and the need to add enzyme right before beginning the series of measurements). Second, it allows you to see what would happen in the later parts of this experiment if you were to use **too much** enzyme: you should see here that with 25ul of **undiluted** enzyme a high A405 value is reached so quickly that it prevents useful data from being collected. Since the kinetics data that we are collecting must be linear over the time period that we are collecting it, this should reinforce that less enzyme will be needed for the kinetics experiment below.

- Be sure to record the color of the mixture when you remove the cuvette from the spectrophotometer. **Remember** from Lab 2A that when a spectrophotometer is measuring high absorption of a particular wavelength, it means we are **not** seeing it! The visible portion of the spectrum spans from approximately 400nm (violet) to 700nm (red), with green-yellow in the 600-500nm range. Compare the shape of the spectrum to the color that you see in the cuvette.

- Important: this shift from the **standard curve** portion to the **kinetics** portion represents a shift from using **PNP** (the colored reaction product) to **PNPP** (the substrate for alkaline phosphatase). Avoid getting these two confused in your notebook!

4. **Perform the actual kinetics experiment, using diluted enzyme.** The entire class will generate **one** dataset that will be used for the kinetics calculations in Lab 4A. Therefore, in coordination with the class and your TA, determine which mixture(s) you and your partner will personally be responsible for generating,

in order to allow the entire class to have one full set of kinetics data by the end of the lab period. Again be sure to use alkaline buffer as the blank.

    a. **This kinetics experiment uses enzyme that has been diluted 4-fold. Do this carefully in alkaline buffer (not water): for example, mix 100ul of enzyme and 300ul of alkaline buffer in a 1.5ul tube, and be sure to label the tube.**

| mixture | volume of substrate (30mM PNPP stock) | volume of alkaline buffer | volume of **diluted** enzyme | final [PNPP] |
|---|---|---|---|---|
| mixture 1 | 5ul | 970ul | 25ul | 0.15mM |
| mixture 2 | 15ul | 960ul | 25ul | 0.45mM |
| mixture 3 | 25ul | 950ul | 25ul | 0.75mM |
| mixture 4 | 50ul | 925ul | 25ul | 1.5mM |
| mixture 5 | 100ul | 875ul | 25ul | 3.0mM |
| mixture 6 | 150ul | 825ul | 25ul | 4.5mM |
| mixture 7 | 225ul | 750ul | 25ul | 6.75mM |
| mixture 8 | 350ul | 625ul | 25ul | 10.5mM |

5. **Record the kinetics results**. The spectrophotometer's kinetics program records the A405 value every 20sec, nine times, after a 3sec delay. Fill in the following table in your notebook **and** transfer the data to the master spreadsheet that the class is collecting for use in Lab 4A.

| mixture | final [PNPP] | time (sec) | A405 |
|---|---|---|---|
| | | 3 | |
| | | 23 | |
| | | 43 | |
| | | 63 | |
| | | 83 | |
| | | 103 | |
| | | 123 | |
| | | 143 | |
| | | 163 | |

    a. Be sure to record the color of the mixture when you remove the cuvette from the spectrophotometer.

    b. If time permits, you should repeat your kinetics measurement with a fresh mixture. It isn't completely essential to replicate your data, **but** the entire class will be using your data in Lab 4A, and it can be difficult to get everything mixed and transferred to the cuvette and promptly measured on the first try! Therefore, if time allows, you can consider your first attempt to be a practice run, and then the second attempt to be your actual experiment.

6.  **Perform additional measurements**. Depending on the size of your lab section, you may find that after the measurements on your assigned mixture(s) have been performed, you still have quite a bit of time remaining. Verify with your TA whether any mixtures still need to be prepared and analyzed on the spectrophotometer in order to complete the group dataset for Lab 4A. If that group dataset is complete, then you have three options for extending your work with the remaining time:

    a.  try to generate your own complete dataset (all eight mixtures) so that you can see whether you can replicate the group's combined work. This also allows you to compare the extent of the color change (i.e. PNP formation) for each mixture by eye as you remove each cuvette from the instrument.

    b.  give someone else a turn at generating a complete set of standard curve data – it's great pipetting practice! In Lab 4A you can see whose $r^2$ value for their standard curve is closest to 1.0, indicating excellent care in the preparation of each mixture.

    c.  designate at least one of the NanoDrop instruments to be for 37C use, locate the software option that allows you to heat the cuvette to 37C for the duration of the kinetics experiments, and repeat your measurements at this temperature to determine whether exposing this mammalian enzyme to its typical cellular temperature influences its reaction rate – make a prediction and test it!

7.  **Share the kinetics results**. Every member of the class will need to have the exact **same** kinetic data for mixtures 1-8 for use in Lab 4A. Verify with your TA whether you will receive the class data spreadsheet by email or via Blackboard. **You will need to have the spreadsheet on your laptop in Lab 4A** (or else plan to use one of the desktop computers in the lab).

8.  **Clean up**. PNP in particular is **toxic**, and needs to be disposed of in a designated waste container in the fume hood. There is also a designated waste container for the empty cuvettes. Spectrophotometers should be turned off (switch is at the back), and the upper and lower parts of the pedestal (the parts that came in contact with the standard curve samples) should be carefully and thoroughly wiped with a water-soaked lint-free tissue and then a dry tissue, so that no sample dries on. Clean up your bench completely.

9.  **Look over the assignment questions**. With any remaining time, look over the assignment questions and the background information about the kinetics parameters ($K_m$, $V_{max}$, etc.) that will be calculated in Lab 4A. Enzyme kinetics is a challenging topic, which is why it is split over two labs (3A and 4A), so it is worthwhile to look ahead to see what questions you may have!

10. **Hand in a copy of your lab notebook pages for grading before you leave at the end of the lab period.**

   a. Data collected today will be included in the Enzyme Kinetics assignment, which will be due later, after Lab 4A is complete. Be sure that you have a written record that will allow you to recapture what you did and observed, so that you can access it when you need it!


References for Lab 3A

This lab procedure was modified from the earlier version of the course manual in order to incorporate use of the NanoDrop instruments.

The image showing the reaction catalyzed by alkaline phosphatase was obtained from the *p*-nitrophenol phosphate purchasing information page for Sigma-Aldrich catalog #P7998 http://www.sigmaaldrich.com/catalog/product/sigma/p7998

# Lab 3B: Creating favorable conditions for protein synthesis



- Review/examine two Nobel Prize-winning discoveries related to the Central Dogma:
    - The genetic code – Nobel Prize in Physiology or Medicine, 1968; more information available at http://www.nobelprize.org/nobel_prizes/medicine/laureates/1968/
    - Ribosome structure – Nobel Prize in Chemistry, 2009; more information available at http://www.nobelprize.org/nobel_prizes/chemistry/laureates/2009/press.html
- Use Jmol to visualize an aminoacyl-tRNA synthetase bound to its cognate tRNA and generate a high-quality image.
- Use Jmol to visualize a ribosome bound to mRNA and three tRNAs, and explore the rRNA and protein content of the ribosome.



- Review the Central Dogma (DNA -> RNA -> protein) via animations, in order gain a sense of the rate of speed and complexity of these central life processes.
- Download and manipulate a threonyl-tRNA synthetase bound to a tRNA. Generate a high-quality image highlighting:
    - Each base of the threonine anticodon
    - The 3' end of the tRNA, at which Thr attachment occurs
    - A bound AMP in the active site
    - The cofactor in the active site
    - The enzyme, showing that it can specifically recognize the correct tRNA by making contact with the anticodon
- Appreciate that aminoacyl tRNA-sythetases, not the ribosome, serve as the actual 'translators' of the genetic code, by being the enzymes actually responsible for creating the correct covalent linkages between a particular tRNA and the corresponding amino acid. Observe a synthetase binding to both the anticodon and the acylation site (i.e. 3' end) of a tRNA simultaneously, and noting where addition of the amino acid will occur.
- Download and manipulate a ribosome in Jmol, focusing on the relative locations of rRNA and proteins, as well as the tRNA and mRNA binding sites.
    - View the peptidyl transferase center, and determine whether it is lined with rRNA or protein, to provide confirmation of whether a ribosome can be characterized as a 'ribozyme'.
    - View the binding of the tRNA anticodons to mRNA in both the P site and A site.

- Use these exercises to gain further insight into textbook Figures 30.11, 30.14, and 30.16, which provide important information but represent highly three-dimensional structures in a flat format.




The Jmol exercises assume that you have Jmol installed on the computer that you are currently using. Exercise 19 also assumes that you recall some of the commands from the earlier exercises from Lab 2B (Exercises 1 and 2). You may need to refer back to the instructions in those exercises in order to understand and input some of the commands in Exercise 19.




This dry lab is intended to be performed during your regular lab period – see the lab schedule for details. You should plan to use the scheduled 2hr 50min lab period to complete the steps listed in this dry lab and **also** complete and submit the corresponding assignment during this time.

1. **View the following animations from the Cold Spring Harbor Laboratory's 3D Animation Library:**
    - "Transcription & Translation: The Central Dogma of Biology": https://www.dnalc.org/resources/3d/central-dogma.html
    - "Translation (Advanced)": https://www.dnalc.org/resources/3d/16-translation-advanced.html

2. **Using Jmol, make your own customized version of textbook Fig. 30.11, showing a threonyl-tRNA synthetase bound to its corresponding tRNA, to AMP, and to its cofactor.** You will individually highlight the three bases in the tRNA's anticodon (bases 34-36), and then the base at the 3' end of the tRNA which would serve as the site for attachment of the Thr (base 76). You will also highlight the bound AMP and the zinc cofactor, and display the enzyme in such a way that its overall shape will be recognizable without obscuring the other features that you are highlighting (tRNA, AMP, cofactor).

    - Familiarize yourself with the names of the colors that are available at http://jmol.sourceforge.net/jscolors/#JavaScript%20colors . You will be customizing your figure with seven color choices that **will be called T, U, V, W, X, Y, and Z below**, so that you can (respectively) color the majority of the tRNA, the 5', middle, and 3' position of the anticodon, the 3' end of the tRNA, the metal cofactor, and the enzyme itself. (And presumably you will leave the background black; otherwise your first step in the instructions below should be the command "**background white**" – or whatever color you choose.) **Make note of all of your color choices (T-Z), because part of the assignment will be to make a detailed figure legend for your figure that you will create, and you will refer to each portion of the figure by the name of the color that you chose.**

- Please proceed with the following steps:

  i.  In the top menu in Jmol go to **File -> Get PDB** and enter the identifier **1QF6** {the file should load right away. If you rotate it around, you can hopefully tell immediately that it contains both nucleic acid and protein components!}

  ii.  As always, go to **File -> Console** to display the script console if it doesn't appear automatically.

  iii.  Type **delete water** into the console.

  iv.  Type **restrict rna**

  v.  Type **wireframe only**

  vi.  Type **wireframe 120**

  vii.  Just for fun, type **color group** – this makes the 5' end of the tRNA blue, gradually changing to red at the 3' end, so that you can more easily follow the path of the backbone. See one floating green monomer not attached to anything else? That's the AMP – we'll work on it later. Next, we are going to re-color the tRNA a uniform color, so that when we highlight the anticodon and the 3'end they will stand out.

  viii.  Type **color T** {where T is the name of the color that you are choosing for the entire tRNA. Make note of your color selections, because you are going to need them in order to create a detailed caption for your image}

  ix.  Type **select 34**

  x.  Type **color U** {where U is the name of whatever color you are choosing for the 5' position of the anticodon}

  xi.  Type **select 35**

  xii.  Type **color V** {where V is the name of whatever color you are choosing for the middle position of the anticodon}

  xiii.  Type **select 36**

  xiv.  Type **color W** {where W is the name of whatever color you are choosing for the 3' position of the anticodon. Your entire anticodon should be nicely color-coded now. Its sequence is 5'-CGU-3'. If you rotate the tRNA around, can you at least tell that it's pyrimidine-purine-pyrimidine?}

  xv.  Type **select 76**

  xvi.  Type **color X** {where X is the name of whatever color you are choosing for the final 3'-most position in the entire tRNA. This is where this enzyme would attach the Thr! Can you figure out exactly where its 3'OH is?}

  xvii.  Type **select AMP**

  xviii.  Type **color cpk**

xix.    Type **spacefill** {this will make this small molecule easier to find when you add in the enzyme. This AMP is 'used' ATP – the portion of ATP that gets covalently added onto an amino acid part-way during the activation/charging process that the synthetase orchestrates. As it turns out, this crystal structure does not actually have the Thr itself, but if Thr were bound, it would be right near here.}

xx.    Type **select Zn**

xxi.    Type **spacefill**

xxii.    Type **color Y** {where Y is the name of whatever color you are choosing for the zinc cofactor}

xxiii.    You're about to add back the enzyme. Before doing so, take a moment to rotate the tRNA around, and try to appreciate why tRNAs are described as being extensively base-paired, with numerous helical regions.

xxiv.    Type **select protein**

xxv.    Type **cartoon** OR type **wireframe** OR type **backbone** OR **strands** OR **spacefill** {you can press the 'undo' button or type undo if you don't like what you just did!}. Choose the depiction of the protein that you think will give you the best view of all the other work that you did!

xxvi.    Type **color Z** {where Z is the name of whatever color you are choosing for the enzyme}

xxvii.    There! You have used the exact same crystal structure data that the textbook authors/illustrators used for Fig. 30.11 (you can tell by looking at their caption), but with the opportunity to highlight key features of the tRNA. In particular, you should have been able to color-code the 3' acceptor stem of the tRNA, make the AMP and cofactor more visible, and color-code each base in the anticodon.

xxviii.    Rotate your molecule around a lot, to get a feel for what you're looking at. When you get a good angle that you like, you can go to **File -> Export -> Export Image** and save a .jpg file for your assignment – or, if you can't find one angle that captures everything then you can save two images that capture completely different angles (probably one that gives the best view of the anticodon and one that gives the best view of the 'acylation site', which is where the amino acid would get transferred from the AMP to the 3' end of the tRNA. If you want one of the views zoomed in, try moving your cursor down while you have the 'shift' key held down – or else **Display -> Zoom** on the menu bar). The assignment will ask you to write a detailed caption to describe what you have created, including what colors and styles you chose.

xxix.    Create a detailed caption for your image, following the instructions in the assignment.

xxx.    Be certain that you are completely finished with your work with this molecule before moving on to the next task.

3.  **Using Jmol, complete <u>EXERCISE 19: RIBOSOMES</u>,** which was modified from the corresponding exercise in "Studying Protein and Nucleic Acid Structure with Jmol". The modified instructions appear here – note that you will have only one

Jmol file open at a time. **Finish up your responses for one Jmol file before opening a new one, so that you don't lose any of your work!**

- To complete **<u>EXERCISE 19: RIBOSOMES</u>**, please proceed with the following steps:

    i.     Go to **File -> Get PDB** and type in **1J5E** – you can see from the console information that it is a 30S ribosomal subunit, and of course you would be able to get further information about this file from the PDB website. For example, the console tells you that there is a chain called 'B', but the website will tell you that 'B' corresponds to the 256-amino acid ribosomal protein S2, and provides links to more information about that protein.
    
    ii.     In the console, type **cartoon only** – it should be easy to tell which color corresponds to RNA and which color corresponds to protein (look for the nitrogenous bases!).
    
    iii.     The small subunit has a fairly flat platform-like shape – as depicted in the textbook on the opening page of Ch.30. Given that rRNA (as opposed to protein) is known to make important contacts with the template and the tRNAs, you should be able to guess which surface of this platform is in contact with the large subunit. The other surface (the protein-rich one) is in contact with the cytoplasm.
    
    iv.     Type **select protein; color chain** – this emphasizes the large number of different proteins that are found in the small subunit (though this structure happens to be missing one of them, known as S1).
    
    v.     Type **spacefill** to see the proteins better. Again rotate the structure to see that there is a protein-rich and a protein-deficient side.
    
    vi.     Type **restrict rna** to view the rRNA alone, and notice that the overall shape of the subunit remains – the central core of the subunit is formed by the rRNA, which is then stabilized by numerous small ribosomal proteins.
    
    vii.     Next let's look at a structure of a large ribosomal subunit. Go to **File -> Get PDB** and type in **1JJ2** – this large structure of more than 90,000 atoms may take a few moments to load. You can see from the console information that it is a large ribosomal subunit, and of course you would again be able to get further information about this file from the PDB website.
    
    viii.     In the console, type **cartoon only** – again, you should be able to locate a relatively flat RNA-rich surface that would contact the other small subunit in a fully-assembled ribosome.
    
    ix.     There are two rRNA molecules here, and in the file the 23S rRNA is coded '0' and the 5S rRNA is coded '9'. Color them separately by typing **select :0; color green** and **select :9; color yellow** – as expected (given then meaning of 'S'), you can see that 23S rRNA is much larger than 5S rRNA.

x.     To see the number and distribution of the ribosomal proteins in this subunit, type **select protein; color chains** and then **spacefill**

xi.     The final file that we'll look at is the most complicated, since it includes a ribosome (large and small subunit) and tRNAs and mRNA – the coming together of three major different types of RNA! It is a complex combined file, and can't be accessed via the File -> Get PDB function. Instead, go to http://www.rcsb.org/pdb/ and type in **4V42** – in the top right click on 'Download Files' and then download the **PDBx/mmCIF File (gz)**. There is no need to expand it; just to in Jmol to **File -> Open** and then locate that file. Note that this structure is missing many atoms in the ribosome, though the tRNAs and mRNA are shown at high resolution.

xii.     Type **select rna; color chain** – the three tRNAs (one each in the E site, P site, and A site) should be evident, and if you rotate the structure you should be able to see a short mRNA template as well. You can tell which tRNA is in the E site, because it is not bound to the template (see textbook Fig. 30.16 and 30.17).

xiii.     To select the tRNAs and mRNA and change their style type **select :AB, :AC, :AD, :A1** and then **spacefill** – it should now be even easier to see that the tRNAs in the A site and the P site are bound to the template (while the tRNA in the E site is not). (AB, AC, and AD are codes in this file for the three tRNAs, while A1 is the code for the template.) Moreover, if you look at the tRNAs that are in the A site and the P site and look at their <u>opposite</u> ends from where the template is, you will be looking at the 'peptidyl transferase center' – the actual site of catalysis in the ribosome, where peptide bond synthesis occurs. This is where transfer of an amino acid onto the growing poly peptide chain occurs – one of the most exciting events in the ribosome!

xiv.     If ribosomes are considered to have 'ribozymal' activity, this means that the RNA component (the rRNA, not the ribosomal proteins) is responsible for catalysis. Can we see this here? Let's look to see what ribosomal components are in the vicinity of the peptidyl transferase center. Type **select protein; spacefill; color grey**

xv.     Now type **select :AA, :BA, :BB** and then **color yellow; spacefill** – as you can probably guess, AA, BA, and BB were the coded names in this file for the three different rRNAs (two in the large subunit and one in the small subunit).

xvi.     Finally, a challenge: type **slab on; slab 50** so that you are looking at a slice through the center of the ribosome, and then rotate it around and see whether you can capture a view that shows the peptidyl transferase center surrounded by yellow (representing rRNA) rather than grey (which represents protein). (How do you recognize the peptidyl transferase center? Remember, you're looking for two template-bound tRNAs, but you're looking at the end of the tRNAs that are <u>farthest</u> from the template, because the end farthest from the template is the end that holds the polypeptide chain and adds a new amino acid to it.)

Remember, if you want to zoom in, hold down the 'shift' key while you click and (while still clicking) move your cursor down.

xvii.   Once you have an angle that you're happy with, take a photo of your screen that shows three things: your view of the peptidyl transferase center (showing evidence that it's a ribozyme), your console, and your student ID in the lower right. This will be part of your assignment submission.

# Lab 4A: Alkaline phosphatase kinetics and active site



Overview of Lab 4A

- Take fourth quiz: kinetics and active site chemistry.
- Perform necessary calculations and graphing in order to obtain estimates for alkaline phosphatase $K_m$, $V_{max}$, and $k_{cat}$.
- Explore the active site residues of alkaline phosphatase using Jmol.
- Use mutagenesis data to infer whether specific residues are involved in binding and/or catalysis.
- Examine the kinetically distinguishable effects of different types of reversible inhibitors, using published data.

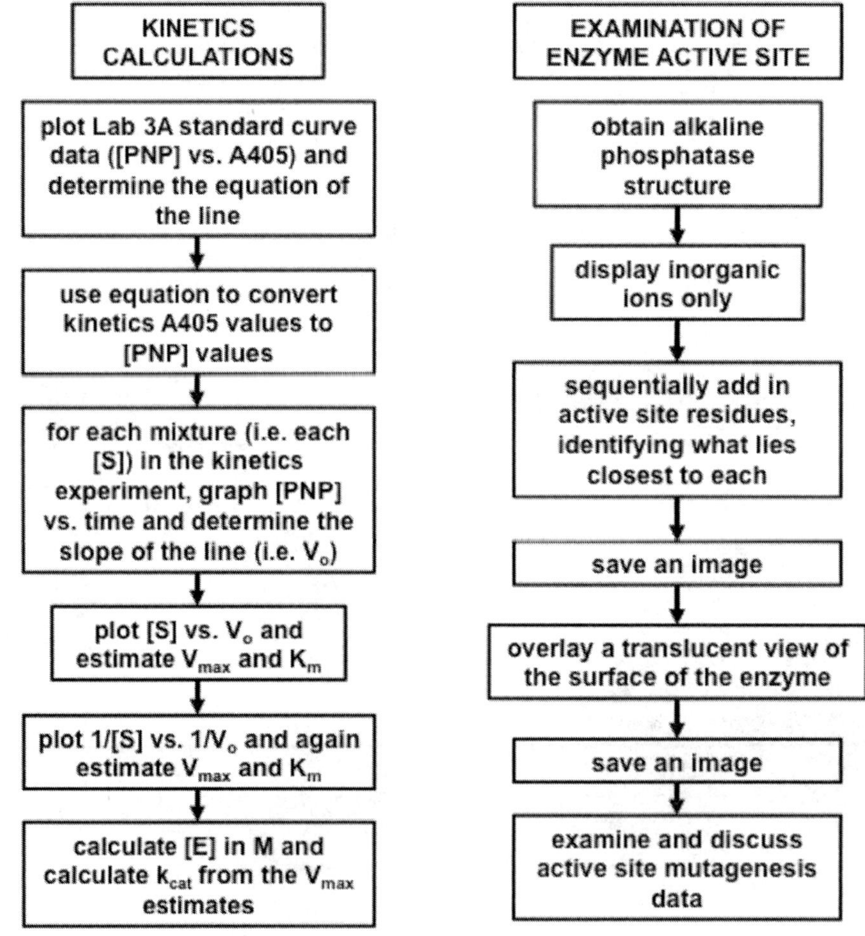

- Complete fourth quiz (**Quiz 4A**) on assigned reading indicated in the 'Preparation for Lab 4A' section.
- Use real data to gain a better understanding of key enzyme parameters and how they are calculated: $V_o$, $V_{max}$, $K_m$, $k_{cat}$.
- Perform a guided in-lab examination of the alkaline phosphatase active site and the likely contributions of specific residues, using Jmol and mutagenesis data. Work to answer the following questions:
  - Alkaline phosphatase requires multiple metal cofactors for its function – which specific residues help to hold the metals in place?
  - Alkaline phosphatase binds specifically to the phosphate portion of phosphate-containing substrates – which specific residues help to bind the phosphate?
  - The catalytic mechanism of alkaline phosphatase involves attack of the substrate's phosphorus atom by a nucleophilic R-group in the enzyme's active site – which specific residue serves as the attacker?
- Use mutagenesis data to more clearly distinguish between the information given by $k_{cat}$ vs. $K_m$ – two often-confused kinetic parameters that give very different information about enzyme functioning.



- Prepare for **Quiz4A** by carefully reading the boxed 'Preparation for Quiz4A' reading that follows.
- Read over the complete procedure for the lab.
- Bring your laptop for this lab if possible; otherwise plan to use one of the desktop computers in the lab for the computational parts of the lab.

pH and amino acid review (textbook Ch.1 and Ch.2):

-the alkaline phosphatase reaction was performed at basic (alkaline) pH – does this mean higher proton concentration or higher hydroxide ion concentration?

-review the amino acid structures and expected charge at pH 7-9 for Ala, Arg, Asn, Asp, Glu, His, and Ser (see textbook section 2.1)

Kinetic parameters:

Velocity ($V_o$): has units of M/s (or M/min), since it is the slope of the line when product formation is monitored over time (should be measured early after substrate and enzyme are mixed!).

$V_{max}$: theoretical possible maximal velocity that can be achieved at a given enzyme concentration; same units as $V_o$. Can be calculated or estimated from a graph, when $V_o$ is known for many different substrate concentrations.

$k_{cat}$: $V_{max}$/[E]. Also known as enzyme turnover number. This is the calculated maximum number of reactions that one copy of the enzyme can catalyze per unit of time, so is maximal velocity divided by enzyme concentration. Usually has units of "per second".

$K_m$: the Michaelis-Menten constant. The substrate concentration at which 50% of $V_{max}$ is achieved, so has units of concentration. The only kinetic parameter that we will consider that has a lower value when enzyme is 'performing well' – a low value indicates that the substrate is binding effectively to the active site even when relatively little is present in the reaction (and conversely, a high value indicates that even though plenty of substrate is present, very little of it is binding effectively to the active site).

1. **Briefly prepare a no-enzyme PNPP mixture and take an initial A405 measurement**. Consult with your TA; likely there will be 1-3 mixtures for the entire class. The purpose is to verify that indeed alkaline phosphatase greatly enhances reaction rate – you should hopefully see that the kinetics data for the no-enzyme mixture is exceptionally different than what was observed over the course of only 163sec in the presence of enzyme!

   - Create a 1ml mixture that consists only of alkaline phosphatase buffer and PNPP. It's important to **omit enzyme**, but aside from that you could make any of the eight mixtures from Lab 3A. For example, the no-enzyme version of Mixture 6 would be 150ul of PNPP stock and 850ul of alkaline buffer, to yield 1ml total.

   - Measure and record the A405 right away. Because we expect this no-enzyme reaction to be so slow, you can take 2ul and use the pedestal option, rather than keeping it in a cuvette for the entire lab period, and then separately measure another 2ul near the end of the lab period. However, since the NanoDrop instruments will not be used for any other purpose during this lab period, it may be more convenient to transfer the mixture to a cuvette and make a custom kinetics program that will take infrequent measurements, for example every 15min for approximately 2.5hrs.)

2. **Perform calculations on data from Lab 3A**. A <u>major</u> objective of this data calculation 'workshop' is that all of you should generate useable graphs and calculations for use in your kinetics report. Students vary widely in how quickly they can perform all of the necessary graphs, calculations, transformations, etc. when starting with the same dataset – if you are finding it easy, then check whether anyone sitting nearby needs help... and if you need help, then be sure to **ask during the lab period**! This time is set aside for you to make sure that you know what to do with regards to kinetics calculations, so make sure to take advantage of it!

- First, plot the standard curve data from Lab 3A, which was **[PNP] vs. A405**, in order to determine the equation of the line that relates these two variables. **Confirm with your TA whether the entire class will be using the same set of standard curve data or whether you will each use your own personal measurements.**
   - Create an 'XY' graph (i.e. a scatterplot), making sure that you have the independent variable on the x-axis and the dependent variable on the y-axis. (**Don't proceed until you are sure which variable goes on which axis!** How do you determine which variable is the dependent one and which is the independent one?)

- o Look for the function that will allow you to add a 'linear trendline'. Look for the 'Chart Layout' menu, or use your Excel version's Help function.
- o Look for 'Trendline Options' that allow you to both display the equation on the graph and display the $R^2$ value on the graph. Remember, the equation of a line is **y = mx + b**, where m is the slope and b is the y-intercept.
  - ▪ In some versions of Excel this involves finding the 'Chart Layout' tab, selecting 'Trendline', choosing a linear trendline, and then clicking on the resulting trendline to select the 'Display equation on chart' and 'Display R-squared value on chart' options.
  - ▪ The equation allows you to calculate the [PNP] for any given A405 value that you may have obtained in other parts of the experiment, at least between the upper and lower [PNP] concentrations that you evaluated. The $R^2$ value comments on how close the points come to connecting in a perfectly straight line (which would be an $R^2$ value of exactly 1.0).
- o In your notebook, note the equation of the line, and rearrange it to a form that allows you to solve for x (which is PNP concentration). This is the overall purpose of this standard curve: to allow you to plug in any A405 value and be able to calculate the corresponding nM PNP.

- Next, **convert all of the A405 values from the kinetics experiment to nM PNP.**
  - o Use the equation that you just generated using the standard curve.
  - o Have all of the A405 values in one column of Excel, and then you can quickly apply the same mathematical transformation to every A405 value, in order to convert each of them to nM PNP.

- Next, **calculate the slope (i.e. $V_o$) associated with each reaction tube** from the kinetics experiment. If there were eight different substrate concentrations, then there will be eight different slopes to calculate.
  - o In the same manner as you did for the standard curve, plot your newly-calculated [PNP] vs. time for each of the different reaction tubes on an 'XY' graph, and then show the equation of the line, which will then reveal the slope (that is '**m**' when the equation of the line is in **y = mx + b** form). As before, make sure that the independent variable is on the x-axis and the dependent variable is on the y-axis.
  - o The slope that you have just calculated is $V_o$! There is one $V_o$ value for each graph. What are the units for $V_o$?
  - o (For your reference, what you just plotted is comparable to one of the lines in the **first** panel of textbook Fig. 8.10. Your data should be linear over the entire ~2.5min range, while the textbook data approaches equilibrium. We would likely have needed to continue our measurements for well over 10min in order to see that tapering-off effect.)

- Prepare to **make the Michaelis-Menten plot** as follows:
  - Make a new table that has two columns: one contains each [S] and the second contains each corresponding $V_o$ value that you just determined. Be sure to include the units!
  - Make a new XY plot of these data. Be sure to label the axes and include units.
  - (For your reference, what you just plotted is comparable to the **last** panel of textbook Fig. 8.10.)
  - Review how to estimate $V_{max}$ and $K_m$ from this graph (this is also shown in textbook Fig. 8.11).

- Prepare to **make the Lineweaver-Burk plot** as follows:
  - Make a new table with values that are the reciprocal of those in the Michaelis-Menten data, so that you will be plotting $1/[S]$ vs. $1/V_o$.
  - Make a new XY plot of these data. Be sure to label the axes and include units.
  - On this graph, the y-intercept is estimated to be $1/V_{max}$. Calculate this first. The slope is estimated to be $K_m/V_{max}$. Calculate this second.
  - (This is also shown in textbook Fig. 8.12.)

- **Estimate $k_{cat}$** using the following information:
  - $V_{max}$ has now been calculated in two ways (above)
  - The final concentration of the enzyme in the reaction mixture was 125ug/ml (because the stock concentration was 5mg/ml and the volume used in the reaction resulted in it being diluted 40x).
  - This enzyme's molecular weight is 57,000 g/mol.

- At minimum, be sure that you have written the following in your notebook: whose standard curve data you used in your calculations, your eight calculated $V_o$ values (including units), your $K_m$ and $V_{max}$ estimates from both the M-M plot and the L-B plot, and your best $k_{cat}$ estimate.

3. **Examine the structure of *E. coli* alkaline phosphatase.**
    i. Open Jmol. If the Script Console is not already open, go in the top menu to **File -> Console**
    ii. Go to **File -> Get PDB** and type in **1ED8**
    iii. Some information about this file should appear in the script console; write it in your lab notebook.
    iv. Type **delete water**
    v. Type **cartoon only**
    vi. Type **color group** – this is the command that colors each chain progressively from blue to red going from the N- to the C-terminus. Notice that it's a dimer (that is why, for example, there are two different green-colored regions). Also rotate it a lot to look at the secondary

structure. It has a very specific folding motif known as a 'Rossman fold', with an interior beta-sheet surrounded by amphipathic alpha-helices – see if you can see the interior beta-sheet.

vii. Type **restrict zn,mg,po4**

viii. Type **spacefill**

ix. Type **color cpk**. Click on the atoms, and their identities will be displayed in the script console. You are looking at the inorganic parts of the structure – three bound metal cofactors that facilitate the reaction, and a phosphate that was crystallized along with the enzyme. (Remember, phosphate is involved in the reaction that this enzyme catalyzes – is phosphate a substrate or product?)

x. Next, you will add specific amino acids from the enzyme active site sequentially back in. Be sure to make a record in your lab notebook of whether each side chain is expected to be charged (at pH ~7-9) and which ion is nearest to it (phosphate? $Mg^{2+}$? $Zn^{2+}$? More than one of these?)

xi. Type **select 322**

xii. Type **wireframe 120**

xiii. Type **color cpk** – then click on it (or hover over it) to verify what amino acid it is. Make a note: is the side chain charged? What ion is it closest to? (Note that we are working on only one monomer, so the active site of the other monomer will still contain only metals and phosphate.)

xiv. Type **select 166**

xv. Type **wireframe 120**

xvi. Type **color cpk** – then click on it to verify what amino acid it is. Make a note: is the side chain expected to be charged at pH 7-9? What ion is it closest to it and what is that ion's expected charge? Is there any electrostatic attraction expected between these two? (If more than one ion is nearby, note that.)

xvii. Type **select 51**

xviii. Type **wireframe 120**

xix. Type **color cpk** – then click on it to verify what amino acid it is. Make a note: is the side chain expected to be charged at pH 7-9? What ion is it closest to it and what is that ion's expected charge? Is there any electrostatic attraction expected between these two? (If more than one ion is nearby, note that.)

xx. Type **select 327**

xxi. Type **wireframe 120**

xxii. Type **color cpk** – then click on it to verify what amino acid it is. Make a note: is the side chain expected to be charged at pH 7-9? What ion is it closest to it and what is that ion's expected charge? Is there any electrostatic attraction expected between these two? (If more than one ion is nearby, note that.)

xxiii. Type **select 412**

xxiv. Type **wireframe 120**

xxv.     Type **color cpk** – then click on it to verify what amino acid it is. Make a note: is the side chain expected to be charged at pH 7-9? What ion is it closest to it and what is that ion's expected charge? Is there any electrostatic attraction expected between these two? (If more than one ion is nearby, note that.)

xxvi.     Type **select 102**

xxvii.     Type **wireframe 120**

xxviii.     Type **color cpk** – then click on it to verify what amino acid it is. Make a note: is the side chain charged? What ion is it closest to? (Note that this residue looks a bit unusual because it was captured in two slightly different conformations, and both are being shown simultaneously.)

xxix.     Zoom in (press the shift key, then move your cursor up), to get a better view of the active site that you were working on. Ignore the other monomer that you weren't working on.

xxx.     Rotate your active site around, trying as best you to get a good angle that clearly shows as many of the active site residues as possible interacting with ions (metals and phosphate). Once you have an angle that you are happy with, go to **File -> Export -> Export Image** and save a .jpg file (**not** a photo or a screenshot) for your Enzyme Kinetics report. Before moving on, make a rough sketch in your notebook that labels each amino acid and inorganic ion that can be seen in this view – **you will need this information** in order to make the caption for your figure for the assignment!

xxxi.     Zoom back out (shift, then move your cursor down). Then, add back the enzyme as follows.

xxxii.     Type **select protein**

xxxiii.     Type **ribbon**. This should look a little overwhelming, so try a different display.

xxxiv.     Type **undo**.

xxxv.     Use the pop-up menu (two-finger click or right-click) and choose **Surfaces -> Solvent-Accessible Surface**

xxxvi.     Choose **Color -> Surfaces ->** {whichever color you choose for the enzyme; don't choose one of the darkest colors if you're leaving the background black}

xxxvii.     Choose **Color -> Surfaces -> Make Translucent**

xxxviii.     Be sure to zoom in and zoom out, to get a good view of the entire protein dimer and the modifications that you have made to one monomer's active site.

xxxix.     Rotate until you get a view that shows the active site that you were working on <u>close to the surface</u>. This is the view that you want – the one where you would be able to visualize a substrate actually being able to <u>access and bind</u> to the active site. If you are having trouble, take a look to see if any of your classmates have found a suitable angle yet, to get an idea of what you're looking for! (Remember the function of this enzyme – it's a fairly **nonspecific** phosphatase, and so we'd predict that the most crucial feature of the substrate would be the phosphate group, while the rest of the substrate seems to be relatively unimportant and so does not likely make close contacts within the enzyme's active site. And in this particular crystal structure, there happens to be no substrate – just a free phosphate.)

xl. Once you have a suitable angle, go to **File -> Export -> Export Image** and save a .jpg file (**not a photo or a screenshot**) for your Enzyme Kinetics report. **Unlike** some previous Jmol exercises, your objective here is to show that you know how to export a high-quality image.

xli. Be sure to write down in your notebook the name of the color that you chose for the enzyme, and any other choices that you made (e.g. if you changed the background color), so that this information can be accurately transferred to your assignment caption.

4. **Study the general mechanism of alkaline phosphatase, and discuss data for specific alkaline phosphatase mutants**.

- The following table contains real published kinetic data for wild-type alkaline phosphatase compared to mutant versions with single-residue substitutions. (The overall principle is similar to the mutagenesis experiment that we will study in textbook Fig. 9.15.) It happens to be from a **bacterial** form of the enzyme (just like the Jmol image that you just created was for a bacterial alkaline phosphatase) so if you find that the kinetic parameters in the wild-type are different from what you just calculated for our **mammalian** enzyme that we used in Lab 3A, there is good reason!

- Look carefully at the data, and determine whether each mutation has a **dramatic** effect on $k_{cat}$, $K_m$, or both. One mutation could certainly affect both kinetic parameters, but focus your attention where you see at least a 50-fold effect or so, and ignore smaller changes. The purpose is to try to differentiate between residues that are overwhelmingly involved in substrate binding vs. catalysis.

    o **The catalytic mechanism involves attack by an R-group that serves as a strong nucleophile.** In lecture, we won't be discussing suitable candidate residues until textbook Section 9.1. Without that information, try to **infer** which R-group could be the nucleophile by focusing on what can be learned from the mutagenesis data!

- Be sure to look back at your notes regarding these residues from your Jmol visualization – was there anything in there to help suggest that any of these residues were involved in **substrate binding vs. catalysis?**

| Mutation | $k_{cat}$ (sec$^{-1}$) | $K_m$ (uM) |
| --- | --- | --- |
| Ser102 → Ala | 0.00134 | 44 |
| Asp327 → Ala | 0.01 | 13900 |
| His412 → Asn | 7 | 210 |
| Wild type | 35 | 8 |

- Make notes based on your observations and discussions today to help you fill in a version of the following table in your notebook, so that you will be prepared to answer the corresponding Enzyme Kinetics assignment questions:

| Mutation | From the kinetic data, determine if this residue is largely involved in catalysis or substrate binding (or both). | What is the specific role of this residue in the mechanism of alkaline phosphatase? Provide your best guess. | What information supports this answer? |
| --- | --- | --- | --- |
| **Ser102** | | | |
| **Asp327** | | | |
| **His412** | | | |

5. **A few minutes before you leave, check on the no-enzyme sample**. Hopefully you saw little change in absorbance in the absence of enzyme – this should support the idea that the enzyme that you experimented with in Lab 3A (and performed kinetics calculations on today) is **spectacularly** well designed for enhancing the rate of dephosphorylation reactions!! Although that was a mammalian alkaline phosphatase while your Jmol and mutagenesis exercises involved a bacterial alkaline phosphatase, they are highly conserved in their mechanistic strategy for speeding up the rate of this type of reaction.

Reference for Lab 4A

Grunwald, S. K., Krueger, K. J. (2008). Improvement of Student Understanding of How Kinetic Data Facilitates the Determination of Amino Acid Catalytic Functions Through an Alkaline Phosphatase Structure/Mechanism Bioinformatics Exercise. Biochemistry & Molecular Biology Education, 36(1), 9-15.

# Lab 4B: Enzyme active site residues and substrate binding; conversion of a zymogen to an active enzyme



- View an animated version of the chymotrypsin mechanism of action.
- View and manipulate a small molecule (inhibitor) in Jmol.
- Examine the structure of an enzyme that is activated by proteolytic cleavage, both before and after its activation, and observe key structural changes that occur during its activation.



- Obtain a better understanding of the chymotrypsin mechanism of action, by first viewing an animation and then exploring the enzyme in detail.
- Download the file for a 3D chemical structure (chymotrypsin inhibitor) and view it in Jmol.
- Create an image of active chymotrypsin that individually highlights the three members of the catalytic triad, the oxyanion hole, and the specificity pocket.
- Highlight these same features in inactive chymotrypsinogen, as a means of understanding why it is not yet active.
- Use these exercises to gain further insight into textbook Figures 9.1 and 9.6-9.10, which provide important information about the chymotrypsin mechanism of action, but can be difficult to understand because they represent highly three-dimensional structures in a flat format.
- Use these exercises to gain further insight into textbook section 10.4, which discusses the proteolytic activation of chymotrypsin from the inactive chymotrypsinogen precursor.



This lab assumes that you have already covered textbook section 9.1 in lecture. If this isn't yet the case, then familiarize yourself with section 9.1, particularly the portions of the textbook that discuss Figures 9.1 and 9.6-9.10, before proceeding.

This dry lab is intended to be performed during your regular lab period – see the lab schedule for details. You should plan to use the scheduled 2hr 50min lab period to complete the steps listed in this dry lab and **also** complete and submit the corresponding assignment during this time.

1.  **Watch an animation of the chymotrypsin mechanism.** There are many, many animations available online that you can search for, since chymotrypsin's 'catalytic triad' mechanism of action is complex and widely studied.
    - One example to try to find is called "Chymotrypsin Mechanism of Action" at https://www.youtube.com/watch?v=IykxtxzSwoA
    - Another option is called "Chymotrypsin Mechanism of Action Movie" at https://www.youtube.com/watch?v=RqZG8JQqonA

2.  **Make your own customized combined version of textbook Fig. 9.6 and 9.10, showing chymotrypsin and highlighting its catalytic triad, oxyanion hole, and specificity pocket.** You will individually highlight the three residues in the catalytic triad (Ser 195, His 57, and Asp 102), the region that stabilizes the transition state (i.e. the 'oxyanion hole'), and the region that specifies this particular protease's preference for cleaving after bulky hydrophobic R-groups (i.e. its 'specificity pocket').

    - Familiarize yourself with the names of the colors that are available at http://jmol.sourceforge.net/jscolors/#JavaScript%20colors . You will be customizing your figure with six color choices that **will be called U, V, W, X, Y, and Z below,** so that you can (respectively) color the majority of the protein, the Ser195, the His57, the Asp102, the oxyanion hole, and the specificity pocket. (And presumably you will leave the background black; otherwise remember to use the command "**background white**" – or whatever color you choose.) **Make note of all of your color choices (U-Z), because part of the assignment will be to make a detailed figure legend for your figure that you will create, and you will refer to each portion of the figure by the name of the color that you chose.**

    - Please proceed with the following steps:

    i.   Go to http://rcsb.org/ and search for **6GCH** – this is a record for chymotrypsin bound to an inhibitor called APF. Scroll down to the 'Small Molecules' section, which gives the full name for this chemical, along with the chemical formula and a long code called an InChI Key (International Chemical Identifier) and some download links.

    ii.  Click on the nearby link '**Download CCD File**'. (Saving the SDF file would be an option as well, but that version is missing the hydrogens from the structure.)

iii.     Open Jmol.

iv.     In the top menu in Jmol go to **File -> Open** and open the APF.cif file that you downloaded using the 'CCD' option. (A .cif file is a Crystallographic Information File.)

v.     Rotate the molecule around to get a good look at why this molecule mimics a small peptide. It can act as an inhibitor of chymotrypsin, a protease that cleaves peptide bonds after bulky hydrophobic R-groups, due to the presence and location of the phenylalanine-like region and peptide bond-like region. However, this molecule is not quite similar enough to a polypeptide to be efficiently cleaved, and instead remains in the active site and blocks the binding of substrate molecules. (Note that it has three green atoms – you can click on them for more information, or else deduce from the full name or the chemical formula for this molecule that these are fluorines.)

vi.     After you have rotated the molecule suitably so that you have an angle that **shows, as best you can, the region that mimics the peptide bond and the phenyl ring** (both in the same view – not two different views), take a photo of your screen that includes your photo ID in the lower right, for use in your report.

vii.     Next you will look at chymotrypsin bound to this APF molecule. But first, type **set pdbAddHydrogens** – this will ask Jmol to show hydrogens in the structure, even though they could not be resolved in the original structure data due to technical limitations. You should receive the response 'pdbAddHydrogens = true'.

viii.     In the top menu in Jmol go to **File -> Get PDB** and enter the identifier **6GCH**

ix.     Type **delete water** into the console.

x.     Rotate the structure until you find the three green fluorines, which are marking the location of the active site.

xi.     Type **spacefill** – you should still be able to easily locate the fluorines of the inhibitor, and therefore locate the active site itself. Notice that hydrogens are shown in white.

xii.     Type **select apf**

xiii.     Type **spacefill off; wireframe 120** – now rotate the enzyme so that you can see what a great fit the inhibitor's phenyl group is for chymotrypsin's specificity pocket. Recall, APF is the same molecule that you were examining earlier, before loading the enzyme structure.

xiv.     Watch what happens as you type **spacefill** – you should again see what a good fit this phenyl group is for the 'specificity pocket' of this enzyme.

xv.     Type **undo**

xvi.     Type **select protein**

xvii.     Then in the pop-up menu (right-click or two-finger click) choose **Surfaces -> Solvent-accessible Surface** – this is one of the best ways to visualize how the substrate (or inhibitor, in this case) would diffuse towards and bind to the active site.

xviii.     Type (or click on) **undo** and you will now mark the catalytic triad residues.

xix.     Type **select protein**

xx.     Type **spacefill off**

xxi.     Think about how you want to represent the main portion of the protein so that you will be able to see the interior of the active site Some reasonable options are 'cartoon only', 'backbone 50 only', 'ribbon only', or 'trace only'. You can try any of these, and 'undo' what you don't like.

**xxii.**     Type **color U** (remember, U is the first color from your chosen color scheme)

xxiii.     Type **select 195**

xxiv.     Type **wireframe 180; color V**

xxv.     Type **select 57**

xxvi.     Type **wireframe 180; color W**

xxvii.     Type **select 102**

xxviii.     Type **wireframe 180; color X**

xxix.     The catalytic triad is now color-coded – note which one is positioned to attack a carbonyl carbon of the inhibitor! (Unfortunately double bonds are not shown, so look for red oxygens as clues to where carbonyl groups might be.) Where is the oxyanion hole? It should be quite near the carbonyl oxygen, as shown in textbook Fig. 9.8. Mark the oxyanion hole region next:

xxx.     Type **select 193**

xxxi.     Type **wireframe 180; color Y**

xxxii.     Where is the specificity pocket? Color it all the same color as follows:

xxxiii.     Type **select 189,190,191,192, 215,216,217,220**

xxxiv.     Type **wireframe 180; color Z**

xxxv.     Rotate your structure around until you get a good angle that as clearly as possible shows the catalytic triad, with one of the inhibitor's carbonyl oxygens in the oxyanion hole between Ser195 and Gly193, and the phenyl group in the specificity pocket. It will be a challenge, since there is a lot to show, but just do your best! Definitely zoom in; if you need to re-center, try pressing the shift key and double-clicking. Take a photo of your work, with your student ID in the lower right corner.

xxxvi.     Because of your nice color-coding, it's easy to lose sight of the chemistry of the members of the catalytic triad. While you're still zoomed in, type **select 195,57,102**

xxxvii.     Type **color cpk** – rotate the structure around, recall the animation that you watched, and try to visualize the aspartate's carboxyl group keeping the histidine's imidazole ring (which you may notice is pictured with too many hydrogens) in position so that it is perfectly placed to act as a base and accept a proton from serine, which in its deprotonated form becomes a potent nucleophile, and ideally placed to attack the carbonyl carbon of a peptide bond (or in this case an inhibitor). This much-admired strategy allows serine proteases to have rate enhancements that are perhaps billions of times the rate of uncatalyzed peptide bond hydrolysis!

xxxviii. Type **undo** to restore your color-coding.

xxxix. Now that you've taken your photo, go back to spacefill and see how the color-coded residues fit into the overall structure of the protein: zoom out, then type **select protein**

xl. Type **spacefill** – if you rotate the protein around you should again be able to see that the specificity pocket and the active site are accessible to the surroundings. This makes sense, because this is a protease, capable of breaking peptide bonds in very long polypeptides – so most of the polypeptide substrate does not enter the enzyme at all. See if you can visualize how a long polypeptide could remain alongside chymotrypsin, with only the region contain a peptide bond and adjacent hydrophobic R-group entering the active site and neighboring specificity pocket.

xli. Be certain that you are completely finished with taking your photo before moving on to the next task, since you will now be deleting the residue-specific coloring, and focusing on the fact that active chymotrypsin is generated by posttranslational modification.

xlii. Type **hide apf**

xliii. Type **color chain** – you should see that chymotrypsin is comprised of three distinctly different chains, even though it is encoded by only one gene.

xliv. Type **spacefill off; cartoon only** – you should now see that one of the three chains is short and lacks discernible secondary structure, while the other two are longer with defined secondary structure.

xlv. Type **select 16,194**

xlvi. Type **wireframe 180; color cpk** – what is now highlighted is a critical electrostatic interaction between the side chain of one residue and the N-terminal end of one of the chains (this is where showing the hydrogens is useful – it is easier to imagine the electrostatic interaction that involves an amino group when the hydrogens are present, rather than when just a lone nitrogen is displayed). Notice how close together these two residues are to each other. This is an important feature in active chymotrypsin.

xlvii. Do any disulfide bridges help to keep these three chains together? Type **select cystine** (note the spelling – a cystine is a covalently-linked cysteine-cysteine bridge).

xlviii. Type **wireframe 120**

xlix. Type **color yellow**

l. There are both intra-chain (within chain) and inter-chain (between chain disulfide bonds present. By rotating the structure, you should now be able to see how the shortest chain is held on to the rest of the enzyme structure.

li. Type **select protein**

lii. Use the pop-up menu and select **Surfaces -> Solvent-accessible surface**

liii.    Rotating the enzyme around, you should see one area where there is no colored surface. This is where the hidden inhibitor is located – even though we can't see the bound molecule anymore (because we used the 'hide' command), we can still see where it is bound.

3. **Create an image showing the inactive precursor of chymotrypsin, chymotrypsinogen.** You will use the same U-Z color-coding as you used for chymotrypsin, and compare how key residues are placed in the inactive precursor, compared with the 'active' chymotrypsinogen that you just looked at. (Even though the previous structure contained a bound inhibitor, it had still undergone all of the posttranslational processing needed to convert it from 'zymogen' form to active form; therefore that structure was referred to as 'active' even though it happened to have an inhibitor bound to it.)

- Please proceed with the following steps:

  i.    In Jmol, go to **File -> Get PDB** and enter **1CHG** to get the structure for chymotrypsinogen, the inactive precursor of chymotrypsin.
  ii.    Notice again that it contains several disulfide bonds, and notice from the file description in the console that it does not have any substrate or inhibitor bound.
  iii.    Type **cartoon only**
  iv.    Type **color chain** – and notice that chymotrypsinogen is composed of only one single chain. Change the color if desired.
  v.    Type **select 16,194**
  vi.    Type **wireframe 180; color cpk** – these are the same residues that you highlighted earlier that were electrostatically attracted to each other, but notice how far apart they are now. Moreover, you may recall that residue 16 had a free amino group – it was the N-terminus of one of the chains of chymotrypsin. By contrast, in the chymotrypsinogen precursor that you are now viewing, residue 16 is in the interior of the single chain, connected to residues 15 and 17 by peptide bonds, and does not have a charged terminus at all.
  vii.    Color the catalytic triad, oxyanion hole, and specificity pocket as you did previously for chymotrypsin (same colors U-Z).
  viii.    Type **select protein**
  ix.    Type **spacefill** – you may be able to see now that although the members of the catalytic triad are adjacent to each other, the specificity pocket is not fully formed, and that the enzyme is not ready to bind substrate. This chymotrypsinogen is inactive (referred to as a zymogen), and does not become activated until a peptide bond is cleaved between residues 15 and 16 (by another protease, trypsin). That one peptide bond breakage is all that it takes for chymotrypsinogen to become catalytically active chymotrypsin; it can then be further processed by another chymotrypsin copy in order to remove a few additional amino acids, create a third chain, and become the mature form of chymotrypsin that we saw in

structure 6GCH.  Take a photo of the inactive chymotrypsinogen for your report, with your student ID in the lower right corner.


Reference for Lab 4B

This exercise was modified from "Studying Protein and Nucleic Acid Structure with Jmol, Exercise 12: The Mechanism of Chymotrypsin." By Jeffrey A. Cohlberg, 2015, W.H. Freeman and Company.

# Lab 5A: SDS-PAGE separation of proteins and concentration determination by A280 method

Overview of Lab 5A

- Take fifth quiz: SDS-PAGE.
- Perform SDS-PAGE (sodium dodecyl sulfate polyacrylamide gel electrophoresis) separation of proteins.
- Discuss the purpose of each of the main components in SDS-PAGE.
- Gather further evidence to determine whether strawberry DNA samples isolated in Lab 1A contained any protein contamination.
- Relate differences in protein migration to inherent properties of each protein.
- Use the A280 method for protein concentration determination, as an alternative to colorimetric (Lowry) method from Lab 2A.

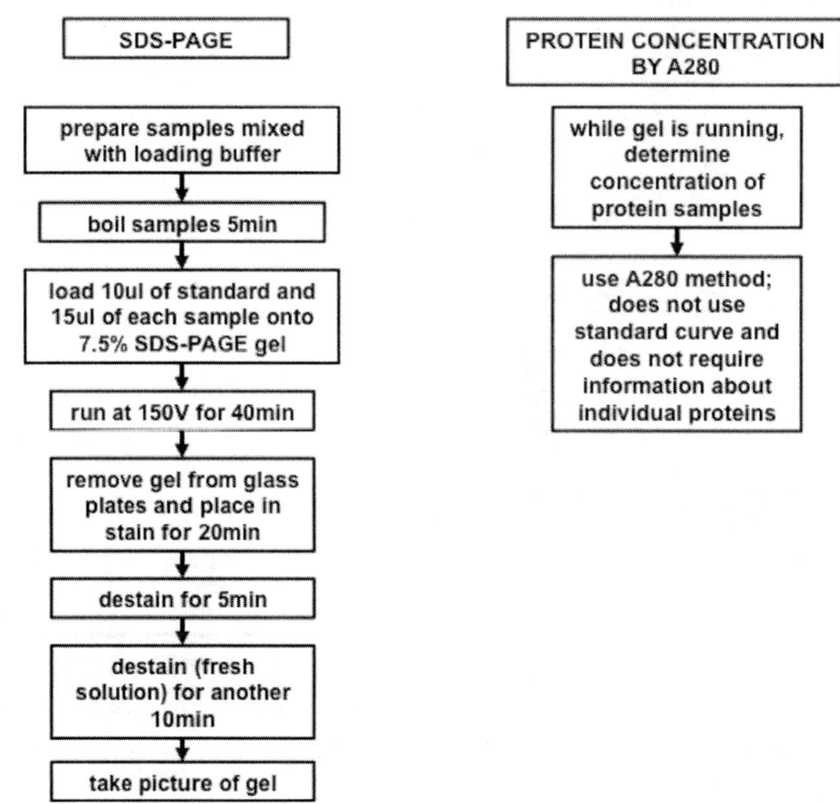

## Objectives of Lab 5A

- Complete fifth quiz (**Quiz 5A**) on assigned reading indicated in the 'Preparation for Lab 5A' section.
- Review the purpose of the major reagents in SDS-PAGE.
- Use SDS-PAGE to evaluate the size of proteins that have been studied in other labs:
    - BSA (bovine serum albumin): used in Lab 2A
    - Alkaline phosphatase: studied in Labs 3A and 4A
    - Chymotrypsin: studied in Lab 4B
    - Malate dehydrogenase: to be studied in Lab 5B
- Use SDS-PAGE to try to determine whether strawberry DNA preparations from Lab 1A contain any detectable protein contamination.
- Observe whether A280-based protein concentration determination correlates with SDS-PAGE band brightness and with protein concentration determinations from the colorimetric Lowry method.


## Preparation for Lab 5A

- Prepare for **Quiz5A** by carefully reading the boxed 'Preparation for Quiz5A' reading that follows.
- Read over the complete procedure for the lab.
- Remind yourself of what a typical absorption spectrum would be expected to look like for DNA (see the **NanoDrop Nucleic Acid Handbook** p.11), since this lab focuses on protein absorption spectra, but at least two of the samples should contain some DNA.
- It is anticipated that you are already familiar with the principles of SDS-PAGE from our Ch.3 coverage. If desired, you may also review a video that discusses this technique at https://www.jove.com/science-education/5058/separating-protein-with-sds-page

**Disulfide bond review (textbook Ch.2):**

-cysteine's side chain ends in a sulfhydryl (-SH) group

-some proteins contain disulfide bridges that covalently link two cysteine residues to each other (-S-S-). This is an oxidized form of the sulfhydryl groups, since electrons are lost during the formation of the disulfide bridge.

-experimentally, these bonds can be broken (i.e. reduced again) in the presence of an excess of a reducing agent that has its own sulfhydryl groups. beta-mercaptoethanol is a popular reducing reagent for this purpose.

-breaking disulfide bonds is desirable in typical 'reducing' SDS-PAGE, when the purpose is to separate polypeptide chains by size, removing any disulfide bridges that hold two chains together or that hold regions of one individual chain together. (It would be possible to perform 'nonreducing' SDS-PAGE, in which the disulfide bonds were kept intact, however migration would no longer be purely based on polypeptide chain length.)

**SDS-PAGE review (textbook Ch.3):**

-PAGE stands for polyacrylamide gel electrophoresis

-sodium dodecyl sulfate (SDS) is an anionic detergent, with a sulfate end (the anion) and a hydrocarbon tail. It serves as a denaturant, disrupting noncovalent interactions within the protein (van der Waals, hydrophobic, H-bond, electrostatic).

-SDS also serves to evenly coat the unfolded proteins with negative charge, masking the inherent charges from the R-groups and instead providing a negative charge that is proportional to their length.

-the gel is composed of polyacrylamide, a polymer of acrylamide. When a current is applied, small proteins (now coated with negative charge) can more more easily move through the polymer towards the positively-charged anode, while larger proteins are more impeded in their migration through the gel matrix, and travel more slowly.

-protein samples are mixed with a loading buffer before being loaded into the gel wells. The loading buffer contains SDS, a reducing agent such as beta-mercaptoethanol, a dye to help track the migration of the sample, and a dense substance such as glycerol to help the sample sink in to the wells as it is loaded.

Procedure for Lab 5A

1. **Prepare the gel apparatus, following the directions of your TA.** A gel box has room to run two gels vertically, and needs to be filled with 1x running buffer (where '1x' means 'at the proper dilution'). The pre-cast gels come with a 'comb' that has ten 'teeth' filling the wells; the comb needs to be removed to create the space for samples to be loaded. It also comes with a piece of 'tape' that needs to be removed from the bottom edge of the gel. The gels that we normally use are 7.5% polyacrylamide.

   - **Chemical hazard! Always use gloves when touching polyacrylamide gels, and do not reuse gloves, since gels may contain traces of toxic unpolymerized acrylamide.**

2. **Prepare the mixtures, by diluting each sample in 2x loading buffer.** (Note: a 2x buffer means that it is twice as concentrated as it needs to be; therefore it should be diluted with an equal volume of protein sample.)

- Gel lanes 1 and 8 will be the molecular weight marker (it does not need to be transferred to a separate 1.5ml tube, and it is purchased pre-prepared with loading buffer included). Gel lanes 9 and 10 are intentionally empty, and can be used if there is a gel-loading error during the loading of lanes 1-8 (of course you would explain this in your notebook).

- Clearly label 1.5ml tubes for the samples below.

| Tube (gel lane) | Sample type | Sample volume | 2x loading buffer volume |
|---|---|---|---|
| 1 | molecular weight marker | 10ul | pre-prepared |
| 2 | bovine serum albumin | 15ul | 15ul |
| 3 | alkaline phosphatase | 15ul | 15ul |
| 4 | chymotrypsin | 15ul | 15ul |
| 5 | malate dehydrogenase | 15ul | 15ul |
| 6 | strawberry DNA (precipitation method) | 15ul | 15ul |
| 7 | strawberry DNA (column purification method) | 15ul | 15ul |
| 8 | molecular weight marker | 10ul | pre-prepared |

- Mix all samples briefly by vortexing. After heating tubes for 5min at 100C in the heat block, briefly (a few seconds) centrifuge all samples to bring all liquid to the bottom of the tubes.

3. **Load 10ul of the molecular weight marker in lane 1, and 15ul of each sample into the appropriate well #2-10.**

- This is great practice with precision pipetting! Make an effort to load 15ul without bubbles, and to avoid getting any samples accidentally mixed into the wrong lanes.

- Be sure to record any mishaps with gel loading in your lab notebook. Use the empty lanes 9-10 as needed. (If needed, lane 8 could also be used for a sample, since it is planned as a duplicate of lane 1.)

4. **Run the gel at 150V for ~40min.**

- If your gel is the only one being run in a particular gel box, be sure to obtain a gel 'dam' from your TA to compensate for the lack of a gel on one side of the apparatus.

- You may be able to notice that the samples first pass through an upper 'stacking' gel, which has a particular composition that yields a compressive effect. Whereas each sample was formerly spread out vertically in a tall well, it begins to compress down into a tight band. After initially passing through the 'stacking' portion of the gel, the sample then passes into the main 'resolving' portion of the gel, where size-based separation occurs.

- Note: it's good practice to visit the gel after ~5-10min and verify that the dye front is migrating as expected, since loose connections and other equipment failures can prevent the gel from running, but can possibly be remedied when detected early!

5. **While the gel is running, assess the protein concentration in each of the stock protein samples, using the spectrophotometer.**

- Rather than performing a full colorimetric assay (which was performed in Lab 2A), this time you will use the 'A280 method'.

- Try the software option that does not use any protein-specific information. In the 'Protein A280' setup option, select "1 Abs = 1 mg/ml" as the type. This option gives the following description: "**A general reference setting based on a 0.1% (1mg/ml) protein solution producing an absorbance at 280nm of 1.0 A.U. (where the pathlength is 10mm or 1cm)**".

- Blank with water, and then record the values below for each of the following samples. Recall that A260 is the peak absorbance for nucleic acids, so a high A260/A280 ratio is indicative of high representation of nucleic acids in the sample. If any of your readings are flagged (e.g. a blue letter 'i'), then tap on that reading to find out why the software is flagging that reading as being of concern, and note the indicated issue in your notebook.

| sample | concentration (mg/ml) | A280 (A.U.) | A260/A280 |
| --- | --- | --- | --- |
| bovine serum albumin | | | |
| alkaline phosphatase | | | |
| chymotrypsin | | | |
| malate dehydrogenase | | | |
| strawberry DNA (precip. method) | | | |
| strawberry DNA (column method) | | | |

- After you have made all six measurements, choose one of the four protein samples (i.e. one out of the first four measurements) and make a quick (~2min) sketch of the absorption spectrum, indicating the approximate peak absorption wavelength. Be sure to take a look at the spectra for the DNA samples as well, and note whether the spectral patterns appear protein-like or not.

- o   Recall that we earlier saw what the expected spectral pattern for **DNA** is – both in the **NanoDrop Nucleic Acid Handbook** (p.11) and in textbook Fig. 4.23. The expected spectrum for a pure **protein** sample looks **different**, with high absorption at A220 (due to the peptide bonds – referred to in textbook Fig. 3.6) and at A280 (due to the side chains of the aromatic amino acids (which ones are they…?), as well as due to disulfide bonds if any are present).  Be sure to note what kind of spectra you are you seeing for your samples today – whether they appear as expected for nucleic acids or for proteins. If neither, then try to think about what types of contaminants or other factors could cause this!

- Note that this is the final opportunity for some good pipetting practice in the final wet lab – make sure that everyone has gotten an opportunity to practice this skill!

- (Also note that there is a different, more customized NanoDrop option that allows the user to enter the extinction coefficient and molecular weight of a particular protein of interest, to allow increased accuracy of concentration estimates.)

6. **When the gel has finished running, stain the gel**.

- **Electrical shock hazard! It is important to turn off the voltage and disconnect the cables before opening the gel tank.**

- With assistance from the TA, pry apart the plastic gel plates, and transfer the gel to the gel stain (Coomassie Brilliant Blue).

- Stain the gel with gentle rocking for 20min.

7. **Destain and photograph the gel**.

- Gently rock the gel for 5min in destain solution.

- Transfer to fresh destain solution, and gently rock for a further 10min.

- Take a photo of your gel as directed by your TA. Be sure that you have a copy of the photo for use in your assignment.

- Check with your TA to verify whether the destain solution will be reused or discarded.

8. **Analyze the gel image**.

- First look at the molecular weight marker (Precision Plus Protein Unstained, from BioRad), and determine which bands are visible (note that the 25, 50, and 75 kDa bands are intended to be darker, to assist in locating them). Note in your lab notebook which bands you are able to see.

    o The following image was obtained from http://www.bio-rad.com/en-us/sku/1610363-precision-plus-protein-unstained-protein-standards-strep-tagged-recombinant-1-ml

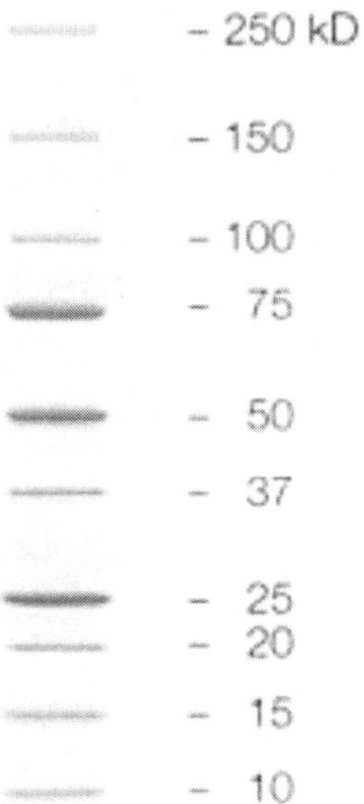


- Next, compare the overall patterns in the four protein samples (lanes 2-5); estimate the apparent sizes of the bands in each of these lanes. (All four of these were purchased as purified proteins, however in some cases only ~85% purity is expected, and so there may be other contaminating proteins present in significant amounts!)

- Next, examine the following information about each of the four proteins, and note whether you see evidence of bands of the expected size. Recall that any disulfide bonds that may be holding quaternary structure together should have been **reduced** by the beta-mercaptoethanol in the loading buffer; in Lab 5B you'll look to see which of these proteins (if any) have disulfide bonds that required reduction in order to allow the protein chains to migrate individually.

| protein | quaternary structure | chain size |
| --- | --- | --- |
| bovine serum albumin | homodimer | 66.5 kDa each |
| alkaline phosphatase | homodimer | 53 kDa each |
| chymotrypsin | 3 different chains | 13.8kDa, 10.1kDa, 1.2kDa |
| malate dehydrogenase | homodimer | 33kDa each |

- Finally, examine the strawberry DNA samples (lanes 6-7) obtained by the precipitation and column purification methods, and note whether they contain any evidence of protein contamination. Does this appear to match the information that you got from the A280 protein concentration determination?

9. **Clean up**. Dispose of the gel destain as directed. Since this is the final wet lab, verify whether you are discarding your DNA samples. Clean your bench area thoroughly. NanoDrop pedestals (both surfaces that come in contact with liquid) should be wiped with a lint-free wipe soaked with water and then a dry wipe, and the instruments should be turned off.

10. **Look over the Protein Concentration Determination and SDS-PAGE Assignment questions.** Use any remaining time to work on questions from the assignment.

11. **Hand in a copy of your lab notebook pages for grading before you leave at the end of the lab period.**

   - Provide your TA with your final set of notebook pages. Be sure that they contain all of the information that you need in order to complete your assignment!

# Lab 5B: Protein disulfide linkages, glycosylation, and the citric acid cycle



- Review/examine a Nobel Prize-winning discovery related to metabolism:
  - **The discovery of the citric acid cycle and of CoA** – Nobel Prize in Physiology or Medicine, 1953; more information available at https://www.nobelprize.org/nobel_prizes/medicine/laureates/1953/press.html
- Create images of the proteins separated by SDS-PAGE in Lab 5A that emphasize any known disulfide bonds and substrate-binding sites.
- Examine a glycosylated protein.
- Examine a protein that catalyzes a redox reaction, and view the proximity of the metabolic intermediate to the electron carrier.



- Appreciate the critical role of the citric acid cycle as the cell's 'metabolic hub' as well as the role of CoA in bringing in each new two-carbon intermediate (i.e. the acetyl portion of acetyl CoA).
- Create one image to illustrate each of the proteins that have been the focus of the previous labs (albumin, alkaline phosphatase, chymotrypsin, and malate dehydrogenase), emphasizing which contain disulfide linkages (both within-chain and between chains) that would have been reduced during the SDS-PAGE procedure in Lab 5A.
- Examine a glycosylated protein, taking a close look at the 3D nature of carbohydrate polymers.
- Examine the interior of a citric acid cycle protein that catalyzes a redox reaction, with particular attention to the proximity of the site of oxidation to the electron acceptor site.



This lab assumes that you have already completed all of the other wet and dry labs for this course. You will need an internet connection for the Jmol exercises; if you would like to work offline, then you should prepare by going ahead of time to http://www.rcsb.org/pdb and downloading the four structure files that you will need (2BXG, 4KJG, 6GCH, and 4WLU), which you can then open with the **File -> Open** command.

This dry lab is intended to be performed during your regular lab period – see the lab schedule for details. You should plan to use the scheduled 2hr 50min lab period to complete the steps listed in this dry lab and **also** complete and submit the corresponding assignment during this time.

1. **Watch an animation of the citric acid cycle.** There are many, many animations available online that you can search for, because of the immense importance of the citric acid cycle (also known as the tricarboxylic acid cycle, or the TCA cycle, or the Krebs cycle). Although textbook Fig. 17.15 contains all of the citric acid cycle steps, it is often useful to view an animation that shows stepwise what is happening to the carbon skeleton, and also that emphasizes where the reactions occur (mitochondrial matrix in eukaryotes). Note that some animations focus specifically on giving you ways to memorize each of the intermediates in order, which will not be our area of emphasis.
   - One example to try to find is an interactive animation at https://www.wiley.com/college/boyer/0470003790/animations/tca/tca.htm
   - Another good option is to look at some videos (non-interactive) from the NDSU Virtual Cell Animations Project. There are several videos relevant to metabolism – search for the NDSU VCell Production videos on YouTube:
     - Citric acid cycle overview: https://www.youtube.com/watch?v=F6vQKrRjQcQ
     - Citric acid cycle reactions: https://www.youtube.com/watch?v=_cXVleFtzeE
     - If you like this format, be sure to go to https://www.youtube.com/user/ndsuvirtualcell/videos to see other comparable animations of relevance to this course, including
       - Glycolysis overview
       - Glycolysis reactions
       - Electron transport chain
       - ATP synthases

2. **Create a figure of human serum albumin, the human equivalent of bovine serum albumin (BSA), emphasizing its role as a carrier protein in blood, and showing its large number of disulfide linkages.** An objective here is to try to further visualize the protein tubes that you used in Lab 5A as more than just small volumes of clear bland liquid – they truly were each distinctive proteins, each with their own highly characteristic structures and functions! Albumin is almost identical in cows and humans, and both contain 17 disulfide linkages per polypeptide. Previously (Lab 2A), you used BSA in order to generate a standard curve for the colorimetric protein concentration assay, since BSA is an inexpensive and very well-characterized protein. Now you will create an image of human serum albumin, and show it functioning as a carrier –

it is the most abundant protein in human blood, and binds to many types of endogenous (i.e. natural) small molecules as well as a variety of medications. In this particular structure you will show albumin bound to the common pain relief medication ibuprofen.

- Familiarize yourself with the names of the colors that are available at http://jmol.sourceforge.net/jscolors/#JavaScript%20colors and choose a color that you would like to serve as your background color. You will be customizing your figures with a background color **X** – you can use the same background color for all five images.

- Please proceed with the following steps:

    i.    Open Jmol, and type **set pdbAddHydrogens** into the console.

    ii.    From the top menu select **File -> Get PDB** and type in **2BXG** – you should be able to see right away that it is a dimer, and maybe even be able to see that it contains a lot of disulfide linkages.

    iii.    Type **select ibp**

    iv.    Type **wireframe 180** – now the bound ibuprofen molecules should be easier to see.

    v.    Type **select protein; trace only**

    vi.    (Optional: you can easily choose a new color Y for your protein here with the command **color Y**)

    vii.    Type **select cystine; color yellow** (note spelling; the cystines are bridged cysteines)

    viii.    Type **set ssbonds backbone** – this connects the disulfide bonds to the backbone, instead of to the invisible space where the cysteine sidechains would end.

    ix.    Make them thicker by typing **ssbonds 180** – if you rotate the image around, you should be able to see that although there are many disulfide bonds (17!) helping to hold the monomeric structures together, there are no disulfide bonds attaching the two monomers to each other.

    x.    Type **select protein.**

    xi.    Use the pop-up menu to choose **Surfaces -> Solvent-accessible Surface**

    xii.    Using the pop-up menu again, select **Color -> Surfaces ->** and then make a color choice, and then **Color -> Surfaces -> Make Translucent**

    xiii.    Recall that you made a choice of background color that here we are calling X, so change the background color with the command **background X** (where X is the name of your color).

    xiv.    Rotate the structure around until you find an angle that shows the ibuprofen clearly – one that shows an ibuprofen molecule as close to the surface of the protein as possible, to help visualize how it would have originally gained access to the binding site in the first place. Focusing on only one monomer is fine.

xv.     Very nice! This image, of the protein that you worked with in Labs 2A and 5A, shows a protein that is typically expected to be found as a dimer, but contains **no** between-monomer covalent linkages; it shows extensive disulfide bridging within each monomer, and that each monomer can serve as a carrier for multiple small molecules (the common pain medication ibuprofen, in this case). On some level it is always hard to believe it, but this truly is the same type of protein that was in the clear, unassuming-looking liquid that you were loading onto your SDS-PAGE gel and taking A280 measurements for in Lab 5A. It should have been in reduced form (no disulfide linkages) while running through the SDS-PAGE gel, and in native (typical, nonreduced) form in the A280 assay.

xvi.     Take a photo of your screen, with your student ID showing in the lower right corner, for use in your report.

3. **Create a figure of a mammalian alkaline phosphatase, emphasizing its disulfide bonds and the fact that it is glycosylated.** Previously (Lab 4A) you examined a bacterial alkaline phosphatase crystal structure, which was an advantage since many mutagenesis studies had been performed on that bacterial enzyme's active site. Now you will examine a rat intestinal alkaline phosphatase, which is highly similar to the bovine intestinal alkaline phosphatase that you experimented with in Lab 3A for the enzyme kinetics lab (since they are both mammalian). Mutagenesis experiments targeting the asparagines at which this protein is N-glycosylated have suggested that glycosylation is important to the function of the mammalian enzyme, although the reason is not completely understood.

- Please proceed with the following steps:

    i.     In Jmol, type **set pdbAddHydrogens** into the console.

    ii.     From the top menu select **File -> Get PDB** and type in **4KJG** – it is hard to tell just by looking at it that this protein is a dimer, however the console indicates this by indicating that there is an 'A' and a 'B' chain. It also indicates that there are two ligands (bound molecules). The first, '4NP', refers to '4-nitrophenyl phosphate', which is another name for the substrate *p*-nitrophenyl phosphate (PNPP) that we used in Lab 3A for the alkaline phosphatase kinetics experiments. The second small molecule is 'NAG', which refers to the modified carbohydrate N-acetyl-D-glucosamine (in our textbook this is depicted in Fig. 11.9 or 11.10, depending on which edition you are looking at). It doesn't mention any metals, but given our work in Lab 4A, we would expect some to be present.

    iii.     Type **delete water**

    iv.     **Type select protein; trace only; color chain**

    v.     Type **select hetero; spacefill** – this view will emphasize the metal ions, the substrate 4NP, and the glycosylation sites. Click on (or hover over) the various structures to see what they are. You should find that, in accordance with what we saw in Lab 4A, the substrate (4NP) is bound near several metals (and that if we explored further, we would uncover appropriately-charged residues that are holding these metals in place, and also holding the substrate's phosphate in place!). You should also find that some instances of the 'NAG' carbohydrate may be small while others may be comprised of several units forming a small polymer. This is fairly difficult to see, so type **select nag** and then **wireframe only; wireframe 120** in order to get a better look at how the carbohydrates. If you zoom in on a NAG monomer, you should see that it is a hexagonal structure, with one oxygen in the ring, a few hydroxyl groups, and some additional substitutions (including one nitrogen). If you zoom in on a NAG polymer, you should notice that it is highly three-dimensional – the inherent bond angles and numbers of functional groups in carbohydrates readily lend themselves to the generation of complex structures.

    vi.    Type **select cystine; color yellow** (note spelling; the cystines are bridged cysteines)

    vii.    Type **set ssbonds backbone** – this connects the disulfide bonds to the backbone, instead of to the invisible space where the cysteine sidechains would end.

    viii.    Make them thicker by typing **ssbonds 180** – you should now be able to see that although this structure contains four disulfide bonds, none of them are connecting the two monomeric chains to each other.

    ix.    The glycosylation sites are still floating, since the asparagine side chains are not shown. Type **select protein.**

    x.    Use the pop-up menu to choose **Surfaces -> Solvent-accessible Surface**

    xi.    Using the pop-up menu again, select **Color -> Surfaces ->** and then make a color choice, and then **Color -> Surfaces -> Make Translucent**

    xii.    Emphasize the glycosylation sites more – type **select nag; spacefill**

    xiii.    Good! You now have a representation of alkaline phosphatase that is very different from the bacterial image that you created in Lab 4A. This one is of the mammalian glycosylated form of the alkaline phosphatase, like the one that you used for the enzyme kinetics experiments in Lab 3A. It shows bound substrate at the active site, bound metals (at the active site as well as additional locations), within-chain disulfide bonds, and several sites of glycosylation. Rotate your enzyme until you get a view that shows bound substrate (remember you can click on it or hover over it to verify that it is 4NP) bound near at least one metal (what kind is it? click on it to verify so that you can mention it in your image caption for your report...), and that shows at least one glycosylation site clearly.

    xiv.    Take a photo of your screen, with your student ID showing in the lower right corner, for use in your report.

    xv.    (Optional ungraded item: if you would like to separately view one of the carbohydrate monomers in Jmol to get a better look, you can search for the 4KJG file at rcsb.org and then near the bottom of the 'Structure Summary' page you can find a link that allows you to download the CCD file for 'NAG' (N-acetylglucosamine), which you can then open in Jmol. This molecule is the same one that is shown in a flat representation in textbook Fig. 11.15.)

4. **Create a figure of chymotrypsin, emphasizing its disulfide bonds and three chains.** Previously (Lab 4B) you examined chymotrypsin's catalytic triad, oxyanion hole, and specificity pocket in depth. Now you will make a quick image that instead focuses on its disulfide bonds and multiple linked chains.

   - Please proceed with the following steps:

     i. In Jmol, type **set pdbAddHydrogens** into the console.
     ii. Go to **File -> Get PDB** and enter **6GCH**
     iii. Type **delete water**
     iv. **Type select protein; trace only**
     v. Type **color chain**
     vi. Type **select cystine; color yellow** (note spelling; the cystines are bridged cysteines)
     vii. Type **set ssbonds backbone** – this connects the disulfide bonds to the backbone, instead of to the invisible space where the cysteine sidechains would end.
     viii. Make them thicker by typing **ssbonds 180** – if you rotate the structure around you should be able to see that there are both within-chain disulfide bonds as well as disulfide bonds that hold the three chains together.
     ix. Type **select protein**
     x. Use the pop-up menu to choose **Surfaces -> Solvent-accessible Surface**
     xi. Using the pop-up menu again, select **Color -> Surfaces ->** and then make a color choice, and then **Color -> Surfaces -> Make Translucent**
     xii. You now have an image that emphasizes the three chains of chymotrypsin and the disulfide bonds that hold these chains together, as well as the inhibitor (which is bound in the specificity pocket and near the oxyanion hole and nucleophilic serine, though these are not highlighted the way that they were in Lab 4B). Rotate the structure around until you find an angle that shows the inhibitor clearly – one that shows the inhibitor as close to the surface of the protein as possible.
     xiii. Take a photo of your screen, with your student ID showing in the lower right corner, for use in your report.

5. **Create a figure of malate dehydrogenase, emphasizing its bound substrate (L-malate) and electron acceptor (NAD+).** Previously (Lab 5A) you examined this protein by SDS-PAGE. Now you will make an image that focuses on the fact that this protein is able to catalyze a redox reaction that serves as the final step in the citric acid cycle, resulting in the oxidation of malate to oxaloacetate. In doing so, it transfers a pair of electrons from malate to NAD+, while

concurrently NAD+ is reduced to NADH. NADH is a crucial source of electrons for the electron transport chain, by which we make most of the ATP that we need in order to survive. Think back to the citric acid cycle animation(s) that you looked at earlier – this enzyme catalyzes the final step in the cycle, allowing the cycle to continue to accept more acetyl CoA.

- Please proceed with the following steps:

    i. In Jmol, type **set pdbAddHydrogens** into the console.

    ii. Go to **File -> Get PDB** and enter **4WLU** – you can see from the text information in the console that this protein contains four chains (A, B, C, and D), and the two bound molecules 'LMR' (L-malate) and NAD+.

    iii. Type **delete water**

    iv. Type **select protein; trace only** – you can get the idea that this protein is a tetramer even without coloring the monomers separately.

    v. Type **color chain**

    vi. It can be determined from the PDB record that this protein does not contain any disulfide linkages (this could be verified by following the steps that we performed previously and trying to show the cystines – we would be unsuccessful).

    vii. Color the solvent-accessible surface as you did previously: type **select protein.**

    viii. Use the pop-up menu to choose **Surfaces -> Solvent-accessible Surface**

    ix. Using the pop-up menu again, select **Color -> Surfaces ->** and then make a color choice, and then **Color -> Surfaces -> Make Translucent**

    x. Rotate the structure around until you find an angle that shows the bound substrate and electron carrier as clearly as possible for one of the monomers – one that helps visualize how they would have gained access to their binding sites in the protein.

    xi. Take a photo of your screen, with your student ID showing in the lower right corner, for use in your report.

    xii. After you have taken and saved a photo, we will make some more changes. The bound malate and NAD+ are still quite difficult to see – this is because they are highly three-dimensional molecules, and are quite deeply buried. After you have taken a photo that contains a view of the protein, type **hide protein**.

    xiii. Use the pop-up menu to select **Surfaces -> Off**

    xiv. We are now going to hide everything except one malate and one NAD; type **hide lmr :A or lmr :B or lmr :C or nad :A or nad :B or nad :C or protein**

    xv. Now what you are viewing is malate and NAD+ in their actual positions relative to each other within a malate dehydrogenase monomer. It only makes sense then that the part of malate that is about to be oxidized is quite close to the part of NAD+ that will receive the electrons. Malate (as shown in textbook Section 17.2) is a 4-carbon metabolic intermediate containing two carboxyl groups and one hydroxyl group. The MDH enzyme's role is to oxidize the hydroxyl group to a keto group, creating oxaloacetate, which is the starting material for a new round of the citric acid cycle. The NAD+, on the other hand, is much larger. Its structure is depicted and discussed in Section 15.4 of our textbook – it consists of ADP (so look for a purine, a ribose, and two phosphates!) attached to another ribose and then a 'nicotinamide' group, which you can recognize as a single nitrogen-substituted aromatic ring attached to an amide group.

xvi. It will be challenging, but rotate the structure around until you get a view that, as best you can, shows malate's hydroxyl group in close proximity to NAD+'s nicotinamide ring. Make sure to zoom in.

xvii. Take a photo of your screen, with your student ID in the lower right corner, for use in your report.

xviii. Finally, type **select lmr or nad; spacefill** – now you can see that those two molecules are so close together that in spacefill view they appear to be just one molecule. This proximity allows the electron transfer to occur, and highlights how well designed the malate dehydrogenase enzyme is. MDH's catalysis of this final step in the citric acid cycle produces essential NADH for the electron transport chain, while regenerating oxaloacetate to keep our indispensable citric acid cycle operational.

xix. (Optional ungraded item: if you would like to separately view the NAD+ molecule alone in Jmol to get a better look, you can search for the 4WLU file at rcsb.org and then near the bottom of the 'Structure Summary' page you can find a link that allows you to download the CCD file for 'NAD', which you can then open in Jmol. This molecule is the same one that is shown in a flat representation in textbook Fig. 15.13, while textbook Table 15.3 tells us that we cannot synthesize NAD+ from scratch, and that one of our reasons for needing to ingest B vitamins (vitamin B$_3$ specifically) is to maintain our supply of NAD+ precursors.)

# Assignments

The lab assignments as currently planned are included in the pages that follow. Any modifications will be announced in advance by your TAs and posted on Blackboard.

**Value: 55pt**
**Late penalty: -10pt per day (i.e. ~18% per day)**

The following eight questions should be submitted in one "DNA Isolation and Analysis" document to Turnitin by the deadline. **The deadlines are strictly upheld, so allow plenty of time for your submission and be certain to save your Turnitin receipt!**

Be <u>certain</u> to include the question number (Q1, Q2, Q3...) along with your response, so that each question can be scored separately as it is graded.

**Q1.** **DNA concentration values, and reproducibility therein (5pt).** Using full sentences, state the concentration values that you obtained for the DNA that **your group personally isolated**, and the number of times that you measured the concentration for the same sample (i.e. whether you had a chance to replicate your concentration measurement). Be sure to state whether you isolated your DNA by the precipitation or the column method, and to report your concentration with suitable units and an appropriate number of significant digits.

- **If you were able to get multiple replicate concentration measurements**, comment on the reproducibility (or lack thereof) in your measurements – if you observed any lack of reproducibility, then are you aware of any specific reasons for the variability in your measurements (such as behaviors that you could correct in the future when using the NanoDrop in future weeks? Were the measurements all made by the same student or did students take turns making measurements)? Choose and defend one DNA concentration value that you are going to use in part 2 below – either why it is reasonable to average them all, or why one or more of the readings should rationally be excluded.

- **If you had only one concentration measurement**, then try to identify and describe the specific reason(s) for why there was not enough time for you to complete replicate measurements. In hindsight, is there anything that you could have done differently that would have sped up the procedure, allowing more time for the concentration measurements?

**Q2.** **Comparison of DNA yield from the two purification methods (10pt).** Calculate the yield for the DNA preparation that **your group personally isolated** as follows: multiply your DNA concentration and the total final volume of DNA (using suitable units!) to obtain the total number of nanograms of DNA isolated. Divide this by the number of grams of strawberry material that you started with in order to obtain a yield in ng/g (i.e. nanograms of DNA per gram of strawberry tissue).

- Explain your calculation in words, in the manner described above, using the specific concentration, volume, and mass numbers from your notebook for this procedure.
- Be sure not to include too many significant digits! Consider both the precision of the NanoDrop and the balance that was used for weighing the strawberries!
- Do a similar calculation for data from a group that performed the other isolation method, so that you will be able to comment on the total amount of DNA isolated by both methods (in ng) and the total yield (in ng/g) by both methods. Then create a table with rows for the two isolation methods and columns for mass of DNA (ng) and for total yield (ng/g).
- Which isolation method gave the higher number of ng of DNA? To what extent can this be explained by the procedures that were used?
- Which isolation method gave the higher ng/g yield? Be **specific** about what you think may have caused one method to have yielded more DNA per gram of starting material (unless they were both equal).
- Having gone through the isolation procedure that **you personally performed**, is there anything that you would do differently next time to try to improve the yield, or would you repeat everything in exactly the same way?

**Q3.** **Analysis of absorption spectra (5pt).** Describe the absorption spectrum for the DNA preparation that **your group personally isolated** first, and for DNA prepared by the **other** method second. (Optional: if it will help you with your discussion, you can include a photo of the sketch from your notebook, but this is not a requirement.) Your submission for this question should include three paragraphs:

a. Compare the absorption spectra to an 'ideal' spectrum as shown in the NanoDrop documentation (p.11 of the NanoDrop Nucleic Acid Handbook). Describe any dissimilarities between the obtained spectra and the 'ideal' spectrum. If your spectrum was flagged by the NanoDrop software as irregular in any way, be sure to expand on that here.

b. Take a look at the absorption spectra for common contaminants during various nucleic acid purification procedures (p.17-18 of the NanoDrop Nucleic Acid booklet). Which wavelengths (approximate ranges) appear to be common 'trouble spots' at which contaminants commonly absorb light? Can you see how looking at full absorption spectra can give enhanced information compared with older spectrophotometric methods that only give single absorbance values at 260nm and 280nm?

c. How likely is it that DNA isolated by either method is contaminated with RNA? (Be sure to look through both isolation procedures for evidence that the sample is being treated with RNAse – if not, then RNA could likely be co-isolated under the same conditions under which DNA is being isolated.) Would you be able to tell by looking at an absorption spectrum whether a

DNA sample was contaminated with RNA? (If DNA and RNA have different peak absorptions, that would certainly be in the NanoDrop Nucleic Acid Handbook!)

**Q4.** **Analysis of NanoDrop output values (10pt).** Reproduce (type out) the data table from your lab notebook, showing the data obtained **by your group personally** first, and by the group that prepared the DNA by the other method second (be sure to clearly label both to indicate how each type of DNA was prepared). Be sure that all five data columns are present for both tables. Your submission for this question should also include three paragraphs:

    a.  Comment on how the A260/A280 and A260/A230 ratios compare with the 'accepted' ratios (see p.14 of the NanoDrop Nucleic Acid Handbook).

    b.  Since the NanoDrop automatically calculates DNA concentration, it is easy to lose sight of the mathematical relationship between absorption and concentration. Consult the "Concentration Calculations" page of the NanoDrop Nucleic Acid Handbook (p.12). In full sentences, restate the relevant equation from this page. (What is concentration directly proportional to, and what is concentration inversely proportional to? What specific extinction coefficient is relevant for DNA?) Also, on the "Concentration Calculations" page, absorbance at both 260nm and 340nm are mentioned. What is the significance of each one? (Refer back to the spectrum on p.11 of the document for confirmation!)

    c.  The molar attenuation coefficient (or 'extinction coefficient') is an inherent property of an analyte. In the case of nucleic acids, if a pure DNA sample and a pure RNA sample had the same A260 reading, which one would have the higher concentration? (Refer back to the "Concentration Calculations" page again.)

**Q5.** **Role of key reagents in DNA precipitation method; comparison with DNeasy method (10pt).** A major purpose of the detergent in the precipitation method is to break up the cell membranes – we'll discuss this later in the course. What is the role of the salt and ethanol? There are many possible sources from which to draw; one suitable one is http://bitesizebio.com/253/the-basics-how-ethanol-precipitation-of-dna-and-rna-works/ but you can use a different one if you prefer.

    a.  Briefly discuss, **in your own words**, the role that the salt and the ethanol play in the DNA precipitation method. Include your citation(s) right at the end of this paragraph. (Incidentally, you might notice your sources mentioning subsequent wash steps that we didn't have time for in our lab – as you can imagine, our lack of wash steps could impact the purity of our final isolated DNA!)

    b.  In addition, discuss the following. Qiagen spin columns have at least two disadvantages compared with 'traditional' DNA isolation methods: relatively high cost, and use of proprietary materials and reagents. Given these disadvantages, what advantages can you think of that might account for the widespread popularity of the DNeasy spin columns in a wide variety of lab settings?

**Q6.** **Analysis of agarose gel photo (10pt).** Include a photo of your gel from Lab 2A, clearly labeling what is in every lane, and clearly labeling the size marker bands.

- Is there visual evidence of any of the purified DNA (by one method? by both?) If so, what size (or size range) does it appear to be?
- Refer back to the DNA concentrations that you used in your calculations earlier in Question Q2. Describe in words how you calculate the mass of DNA that was loaded when you know the concentration and the volume – include the specific numbers that you are using in your calculations. Be sure to do this calculation for the DNA that was isolated by **each** of the two methods – so you should be reporting two volumes, and noting which is for which isolation method. Be clear which lanes you are referring to (e.g. that although there are many samples on the gel, you are going to be focusing on lanes X and Y, which are the ones that you have DNA concentration data for).
- Was there a positive control DNA sample? If so, what size (or size range) did the DNA appear to be? (Be sure to use the size ladder to help you to estimate the range of DNA sizes in your sample.) Again calculate how much DNA was loaded into that gel, using the concentration and loaded volume data from your notebook.

**Q7.** **Three recommendations for the future (5pt).** In the 'real world' researchers generally have many chances to practice and refine their techniques and procedures. If you had to repeat this experiment, what **three** things would you focus on doing differently?

- Focus on things that can be achieved in the time that we have available and using equipment that we already have.
- You may have other ideas, but consider these options:
  - Do you think that there was a way to break open the strawberry cells more efficiently?
  - Do you think that your DNA was at a suitable concentration to allow you to work with it effectively – or should it perhaps have been resuspended/eluted in a smaller/larger volume?
  - Do you think that you had enough replicates of the DNA concentration in order for you to know whether you were getting an accurate reading?
  - Did you need more practice pipetting small volumes before moving on to using the NanoDrop? Did you need more practice using the NanoDrop itself?

- o Do you think that there is a step that you could have performed more quickly or with less error (e.g. where you lost time or spilled sample, but could improve on this next time)?

**Value: 60pt**
**Late penalty: -10pt per day (i.e. ~17% per day)**

The following eight questions should be submitted in one "Protein concentration and SDS-PAGE" document to Turnitin by the deadline. **Be <u>certain</u> to include the question number (Q1, Q2, Q3…) along with your response, so that each question can be scored separately as it is graded**.

**Q1.** **Absorption spectrum for Lowry assay (10pt).** Describe the absorption spectrum that you obtained in the Lowry assay, based on the one representative sketch that you made. (Optional: if it will help you with your discussion, you can include a photo of the sketch from your notebook, but this is not a requirement.) In your description, include the following:
- Which sample was being assayed here (which tube number it was, and what sample that corresponded to).
- The actual observed color of the sample (what you saw by eye).
- The wavelength that was used in the calculations, and the wavelength that the instrument considered to be 'background' based on default settings for this procedure (see the Handbook, p.13).
- Whether the overall spectral pattern was similar to the one shown on p.13 of the Handbook.
- Your assessment of whether those wavelengths used in the calculations (the one that indicates that protein is present, and the one that indicates only background – no protein) were appropriate. Did you indeed observe high and nearly-undetectable absorbance, respectively, at those two wavelengths when protein was present?
- The range of wavelengths that you estimate to be detectable above the background. (In other words, what <u>other</u> wavelengths could the software have potentially used in order to perform the protein concentration calculations?)

**Q2.** **Lowry raw data (5pt).** Provide a clearly-organized table that contains all of the numerical data that you collected from the NanoDrop for the 12 tubes that were set up. If you have any missing data, clearly explain the reason.

**Q3.** **Standard curve and calculation of unknowns in Lowry assay (10pt).** Provide the following:
- a. A graph of the standard curve, with a regression line through the data and the $r^2$ value indicated on the graph. (If you are unsure of how to do this, note that it will be performed in lab for certain kinetics graphs during Lab 4A, before the due date for this assignment question.)

b. A paragraph that discusses whether or not all three concentrations of your unknown (one undiluted and two diluted) fell within the range of absorbance readings over which the standard curve is valid.
c. A clear and specific statement of your best numerical estimate of what the actual concentration of the **undiluted unknown** was (remember to factor in that in some cases your unknown had been diluted!). Be sure to **explain** how you arrived at this estimate, making specific reference to your numerical data.

**Q4.** **Three recommendations for the future in Lowry assay (5pt).** If you had to repeat this experiment, what three things would you focus on doing differently?
- Focus on things that can be achieved in the time that we have available and using equipment that we already have.
- You may have other ideas, but consider these options:
    o Did you have enough time to incubate your mixtures for the recommended amounts of time? (If not, can you think of a way that you could have saved time in another part of the lab procedure?)
    o Do you think that you were able to get a reliable standard curve without performing replicate readings on any of your tubes?
    o Do you think that you did enough replication in order to get reliable readings for the unknowns?
    o Do you think that it would have been a better learning experience if the students had prepared all of the protein dilutions for the standard curve?
    o Did any of your spectra have unexpected shapes, and if so then can you think of any ways to address this in the future?

**Q5.** **SDS-PAGE gel image (10pt).** The response to this question should include two components:
a. The image of your SDS-PAGE gel.
b. A paragraph describing what samples were run in the gel, and under what conditions the gel was run. This is **not** the results interpretation – it simply explains captions your image by explaining what type of gel was run, and what samples were run. Be sure to include the following:
    o The % of polyacrylamide.
    o How long the gel was run for and at what voltage.
    o Chemical contents of the sample loading buffer.
    o Which samples are found in which lanes, and the heat treatment that the samples received before being loaded.
    o The name and source of the molecular weight marker.

**Q6.** **SDS-PAGE gel analysis (10pt).** The response to this question should discuss each of the five sample types individually, separated into five different short paragraphs as follows:

    a. BSA: did the sample migrate as expected? Make specific reference to the bands, as compared with the molecular size marker.

    b. alkaline phosphatase: did the sample migrate as expected? Make specific reference to the bands, as compared with the molecular size marker.

    c. chymotrypsin: did the sample migrate as expected? Make specific reference to the bands, as compared with the molecular size marker.

    d. malate dehydrogenase: did the sample migrate as expected? Make specific reference to the bands, as compared with the molecular size marker.

    e. Isolated DNA from Lab 1A: did the samples appear to contain any protein contamination? If so, estimate size(s).

    f. If there were problems in the running of your SDS-PAGE gel, discuss them here in a final paragraph – what should be done differently in the future?

**Q7.** **Protein concentration by A280 method (10pt).** Based on the calculated protein concentration for each sample, and the volume that was loaded into the gel, indicate the calculated mass of each protein that was loaded into each gel well. Include the following three components:

    a. Create a data table that includes the following columns, based on NanoDrop measurements (keep in mind that mg/ml and ug/ul are equivalent units!):
- Sample name (give all six)
- A280
- A260/A280
- Concentration (ug/ul)
- Volume loaded on SDS-PAGE gel (ul; review the procedure and be sure that you are reporting only sample volume here, not sample + loading buffer!)
- Mass loaded on SDS-PAGE gel (**calculate this** using other columns in this table; be sure to include units!)

    b. Include a short paragraph about your judgment regarding whether the calculated mass of protein correlated with overall band brightness in each well. (e.g. if one type of sample stained darker than the rest, did you also see a higher calculated concentration for this sample, or not?)

    c. Three methods have now been used to assess whether the two DNA isolation methods from Lab 1A allow protein contamination of the DNA samples (SDS-PAGE, Lowry assay, and A280 assay). Based on these three lines of evidence, make an overall concluding statement (or short paragraph) regarding the two methods with regards to whether they permit contaminating proteins into the final DNA sample. (Do both isolation methods show signs of protein contamination? Only one? Neither? And do the methods of determining protein contamination agree with each other or not?)

The following eight questions should be submitted in one "Enzyme Kinetics" document to Turnitin by the deadline. **Be <u>certain</u> to include the question number (Q1, Q2, Q3…) along with your response, so that each question can be scored separately as it is graded**.

**Q1.** **Standard curve (10pt)**. Plot a graph of [PNP] vs. A405 for **<u>your</u>** standard curve that **<u>your group</u>** generated in Lab 3A (**even if** a different group's standard curve data was used in subsequent kinetics calculations), then caption your graph.
- Include an 'XY' plot ('scatterplot') – no other type of graph (bar graph, etc.) is suitable for this type of data in which you are showing how the dependent variable changes as you change the independent variable. Label the axes and include units.
- Include a trendline and $R^2$ value on your graph. Doing so should have been covered during Lab 4A, and you can also use Excel's 'Help' function.
- Write a full paragraph that serves as a detailed caption, describing what can be seen in the graph (be sure to include units!). It should also indicate what the color of the solution was by visual inspection, and should comment on how close the data points come to approximating a straight line (refer to the $R^2$ value!). Discuss any reasons that could have resulted in a deviation from an $R^2$ value of 1.0 in the case of your group's standard curve setup in particular. Assuming that the [PNP] dilutions were prepared as accurately as possible before the lab began, what do you think that your group could do in the future to improve your $R^2$ value if you were to set up this standard curve again?

**Q2.** **Michaelis-Menten data (10pt)**. Provide the Michaelis-Menten data table and graph that you prepared in Lab 4A, and then caption your graph.
- Provide the data table that you are using to generate your Michaelis-Menten plot. It should have a column for substrate concentration and a column for the corresponding calculated Vo, with units clearly indicated.
- Provide a Michaelis-Menten plot of your data. This is again an 'XY' plot, but hopefully the relationship should not be linear this time! No trendline or $R^2$ value is needed (**do not** add e.g. a logarithmic trendline). Note that textbook Fig. 8.10 depicts the conversion of raw experimental data into a Michaelis-Menten plot – it is comparable to what you are doing here.
- Write a full paragraph that serves as a detailed caption, describing what can be seen in the graph (be sure to include units!). Comment on the extent to which your data does (or does not) fit the pattern expected for an enzyme that displays Michaelis-Menten kinetics (see Fig. 8.11 of the textbook). Provide your **best numerical estimate** of the enzyme's $V_{max}$ and $K_m$ using your plot (don't forget to include units!) – again, these are depicted in textbook Fig. 8.11. Then finally, convert your $V_{max}$ value to $k_{cat}$, using the

equation $k_{cat}$ = $V_{max}$ /[E]. Be sure to be clear in explaining how you are calculating $k_{cat}$.

**Q3.** **Lineweaver-Burk data (10pt)**. Provide the Lineweaver-Burk data table and graph that you prepared in Lab 4A, and then caption your graph. Remember that the purpose of the data transformation is to make the estimate of $K_m$ and $V_{max}$ easier than it was for the Michaelis-Menten plot.

- Provide the data table that you are using to generate your Lineweaver-Burk plot. It should simply be the reciprocal of all values that were used for the Michaelis-Menten plot. Remember that the titles (and units!) of your data columns should reflect the fact that you are taking reciprocals of your two kinetic parameters of interest.
- Provide a Lineweaver-Burk plot of your data. This is again an 'XY' plot – let's see if it's linear! A trendline and an $R^2$ value is useful here.
- Write a full paragraph that serves as a detailed caption, describing what can be seen in the graph (be sure to include units!). Comment on the extent to which your data does (or does not) fit the pattern expected for a Lineweaver-Burk plot for an enzyme that displays Michaelis-Menten kinetics (see Fig. 8.12 of the textbook). Provide your best estimate of the enzyme's $V_{max}$ and $K_m$ using this new plot (don't forget to include units!) – and then make a comparison to the values that you provided in the previous question. Do you agree that you could more readily get an estimate of the enzyme's $V_{max}$ and $K_m$ using the Lineweaver-Burk plot rather than the Michaelis-Menten plot? Finally, provide a statement indicating the $k_{cat}$ value (with units!) that corresponds to this new $V_{max}$ estimate, being clear about how you calculated it.

**Q4.** **Alkaline phosphatase active site figures (10pt)**. Provide the two .jpg figures (**not photos or screenshots**) that you prepared using Jmol, and provide suitable detailed captions.

- Figure 1 should be your image of the active site R-groups and inorganic ions. Your caption should give the following information in complete sentences (you choose the order; think about how to construct it in order to make it as readable as possible):
    - Name of the program that you used to prepare the image.
    - Filename, resolution, and method that was used to determine the structure (method is available within this structure's record at http://www.rcsb.org/pdb ).
    - The name of the organism, the name of the protein, the fact that it is a dimer, and the length of each monomer (look this up when you are looking up the method by which the structure was obtained).
    - The full list of inorganic ions that are in the active site in that structure, and then the subset that can actually be seen in the view that you are presenting.

- o The full list of residues (amino acid type and position number) that you chose to display in the active site (the complete list that we 'unhid'), and then the subset that can actually be seen in the view that you are presenting. Putting labels on the actual figure is optional.
  - o The electrostatic attractions between amino acid side chains and inorganic ions that you predict are occurring in the active site (whether or not they can be seen in the view that you chose to include here). The metal cofactors are necessary for function, and the phosphate is the only part of the substrate that the enzyme is highly attracted to – try to coherently describe how these specific evolutionarily-conserved active site residues appear to be important for the function of this enzyme.
  - o If the figure was constructed with the help of a classmate, give that person's name.
  - o (If the second active site is visible, note that there are two active sites in this dimer, but that the viewer should pay attention to the monomer that is surrounded by amino acids.)
- Figure 2 should have a much shorter caption. Indicate that this is a second view of the same enzyme with the same active site residues highlighted, but indicate which options (style and color) were used for displaying the protein. If you can see that the active site is close to the surface and would be accessible to substrate, be sure to describe this.

**Q5.** **Alkaline phosphatase mutagenesis experiments (10pt).** Discuss the answers to the following questions: *Given the experimental mutagenesis data provided, which residue appears to be most important in the alkaline phosphatase mechanism of action and how do you know? If you had to add one additional residue to your Jmol figure showing additional R-groups that are important for holding the metals in position, which residue could you add in to your figure (and expect to see close proximity to one or more of the metals), and what is its R-group chemistry? If you had to design a mutagenesis experiment to try to prove that this residue was indeed important for metal binding, what would be your basis for choosing your selected mutation?*

- It is impossible to fully describe this enzyme's mechanism of action with the information given. A helpful hint is that some entity serves as a **strong nucleophile** in an attack on the phosphate that initiates the entire reaction sequence. This attacker turns out to be an R-group in the active site; given the mutagenesis data, **which residue in the active site is most likely to play this crucial catalytic role? How can the others be ruled out?** Which mutagenesis data column (impact on $k_{cat}$ or impact on $K_m$) most directly gives us clues about key residues involved in catalysis? What does the other data column tell us about?
- To get more information about metal binding sites within alkaline phosphatase, visit the record for the alkaline phosphatase structure 1ED8 at http://www.rcsb.org/pdb and click on the 'Sequence' tab at the top. This

'Sequence Chain View' display is confusing-looking at first, because it is very information-rich – it contains the **1-letter amino acid sequence** for the *E. coli* alkaline phosphatase protein that we have been studying (split up over several lines), along with many annotation tracks (similar in idea to the Genome Browser annotation tracks that you saw in Lab 1B). Scroll down to the 'Legend' section, which explains what symbols are being used to mark up the protein sequence. There is still a lot of information, so focus on the first legend ('Protein Modification Legend'). Familiarize yourself with the symbols being used to indicate what residues most directly bind to the two different zinc ions and the magnesium ion. Then look for their occurrences **above** specific amino acid locations, and note the amino acid position number and type (there are many places where you can find out what amino acid is represented by a particular 1-letter code, including Table 2.2 of our textbook and many places online). For example, the first residue is T (threonine) and the second residue is P (proline). The first fifty residues in the protein don't bind to any metals, but there is a metal binding site shortly after that.

- Look for all of the residues that are believed to be involved in binding to these metals. Make note of them, and compare that list to the six that we already identified in Lab 4A in the Jmol instructions. Imagine that you were responsible for improving your Jmol image of the residues that hold alkaline phosphatase's residues in place by highlighting **additional** residues in the metal-binding region. Find one amino acids (its name and position number) that you **didn't** highlight previously in your Jmol image but that you would predict that you would find located close to the magnesium or zinc. For example, glutamate 322 is shown, based on its yellow and pink triangles, to be binding to both zinc and magnesium, but it is not a good choice for this assignment, since it is **already** a part of the figure that you created in Lab 4A. (Why are long lines connecting residue 322 and residue 51? To emphasize that these residues are close to each other in 3D space, and that those residues are all contacting the **same** exact ion, not simply different occurrences of the same type of ion. You already saw this phenomenon in action within Jmol.)
- Your goal is to choose a **new** amino acid that you would recommend including in the Jmol visualization exercise in Lab 4A, because you strongly predict that it would lie near at least one of the metals. For this amino acid, describe its chemistry, including its expected charge at pH 7-9, and whether you expect that the side chain would be electrostatically attracted to the metal ion.
- For your chosen amino acid, propose a substitution mutation that you would expect would diminish metal binding, and what your rationale was for choosing that new amino acid. There is no one right answer for a 'correct' mutagenesis experimental design, but you can get an idea of a possible strategy by seeing what substitution

mutations were made in the experiment described in Lab 4A and in
the catalytic triad mutagenesis experiment described in textbook Fig.
9.15. Propose a mutation that would lead to dramatically different R-
group chemistry, and explain why you think that the new amino acid
side chain would no longer be attracted to the metal.

**Q6.** **Alkaline phosphatase inhibition data analysis (10pt)**. In two paragraphs
(one per inhibitor graph), discuss the following enzyme inhibition data.

a. Examine the Lineweaver-Burk plot shown below, which shows the same
type of alkaline phosphatase kinetics experiment that you performed in
Lab 3A (substrate again is PNPP). '**Control**' refers to the standard assay
(enzyme, substrate and buffer are present; no inhibitor), while the
second line indicates a comparable experiment in which an inhibitor was
added – 10mM **L-phenylalanine** (L-Phe). Use the information in your
textbook (Figs. 8.19-8.21) to determine which type of inhibitor
(competitive, uncompetitive, or noncompetitive) L-Phe appears to
represent, based on these data. Discuss how this type of inhibitor differs
from the other two types of reversible inhibitors. (Note that we do not
have enough information to speculate on <u>why</u> L-Phe should serve as an
inhibitor of this enzyme.)
   - For the first graph, be sure that your paragraph includes the
following information:
     o Name of the inhibitor (chemical name).
     o Effect that the inhibitor had on the slope and/or y-intercept
on the Lineweaver-Burk plot.
     o Your assessment of which type of reversible inhibitor L-Phe
appears to be (competitive, uncompetitive, or noncompetive),
<u>and</u> your clear explanation, in your own words, of what
distinguishes this type of inhibitor from the other two. If you
are using our textbook as your resource, you can just explain
that and not give a citation; otherwise you should cite your
source(s).
     o Your judgment regarding whether the inhibitor's effect could
likely have been as readily observed if the data had been
plotted on a standard Michaelis-Menten plot (the relevant
textbook figure is one of Figs. 8-16-8.18, depending on which
type of inhibitor you believe this to be).
     o Your best prediction regarding where L-Phe might bind
within alkaline phosphatase, given the type of inhibitor that it
appears to be. (Consult Fig. 8.14 of the textbook, and also
think back to the figures that you made of alkaline
phosphatase and what you now know about its active site.)

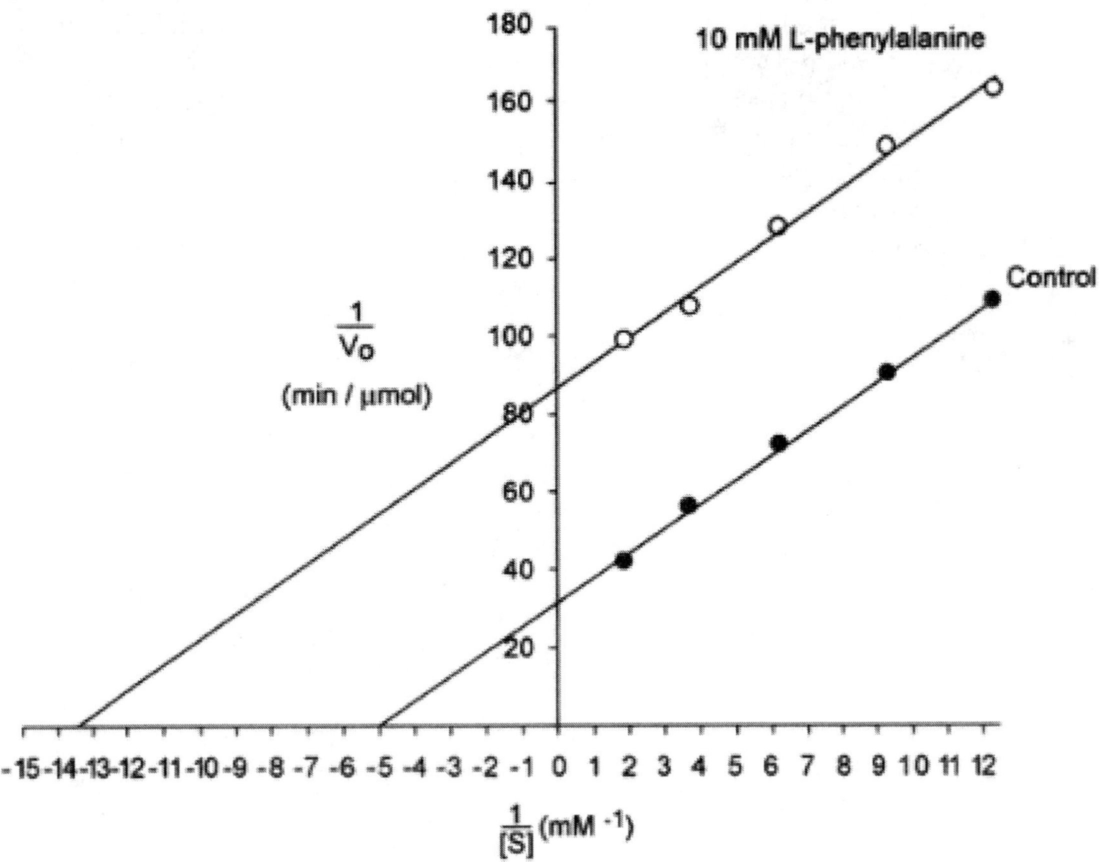
180
160
140
120
100
80
60
40
20
1/Vo
(min / µmol)
10 mM L-phenylalanine
Control
-15 -14 -13 -12 -11 -10 -9 -8 -7 -6 -5 -4 -3 -2 -1 0 1 2 3 4 5 6 7 8 9 10 11 12
1/[S] (mM -1)

b.  Examine the second Lineweaver-Burk plot, shown below, which displays the results of another alkaline phosphatase kinetics experiment (substrate again is PNPP). '**Control**' again refers to the standard assay without inhibitor, while the second line indicates a comparable experiment in which an inhibitor was added – 1mM **phosphate**. Use the information in your textbook (Figs. 8.19-8.21) to determine which type of inhibitor (competitive, uncompetitive, or noncompetitive) inorganic phosphate appears to represent, based on these data. For the assignment, discuss how this type of inhibitor differs from the other two. Also identify the role of free phosphate in the overall reaction normally catalyzed by this enzyme (reactant? product?) and speculate on why phosphate would serve as an inhibitor of this particular type.

- For the second graph, be sure that your paragraph includes the following information:
    - Name of the inhibitor (chemical name).
    - Effect that the inhibitor had on the slope and/or y-intercept on the Lineweaver-Burk plot.
    - Your assessment of which type of reversible inhibitor $P_i$ appears to be (competitive, uncompetitive, or noncompetive), <u>and</u> your explanation, in your own words, of what distinguishes this type of inhibitor from the other two. If you are using our textbook as your resource, you can just explain that and not give a citation; otherwise you should cite your source(s).
    - Your judgment regarding whether the inhibitor's effect could likely have been as readily observed if the data had been plotted on a standard Michaelis-Menten plot (the relevant textbook figure is one of Figs. 8-16-8.18, depending on which type of inhibitor you believe this to be).
    - Your best prediction regarding where $P_i$ might bind within alkaline phosphatase. (Consult Fig. 8.14 of the textbook, and also think back to the figures that you made of alkaline phosphatase and what you now know about its active site.)
- Your educated speculation regarding why $P_i$ might serve as an inhibitor of this enzyme at all.

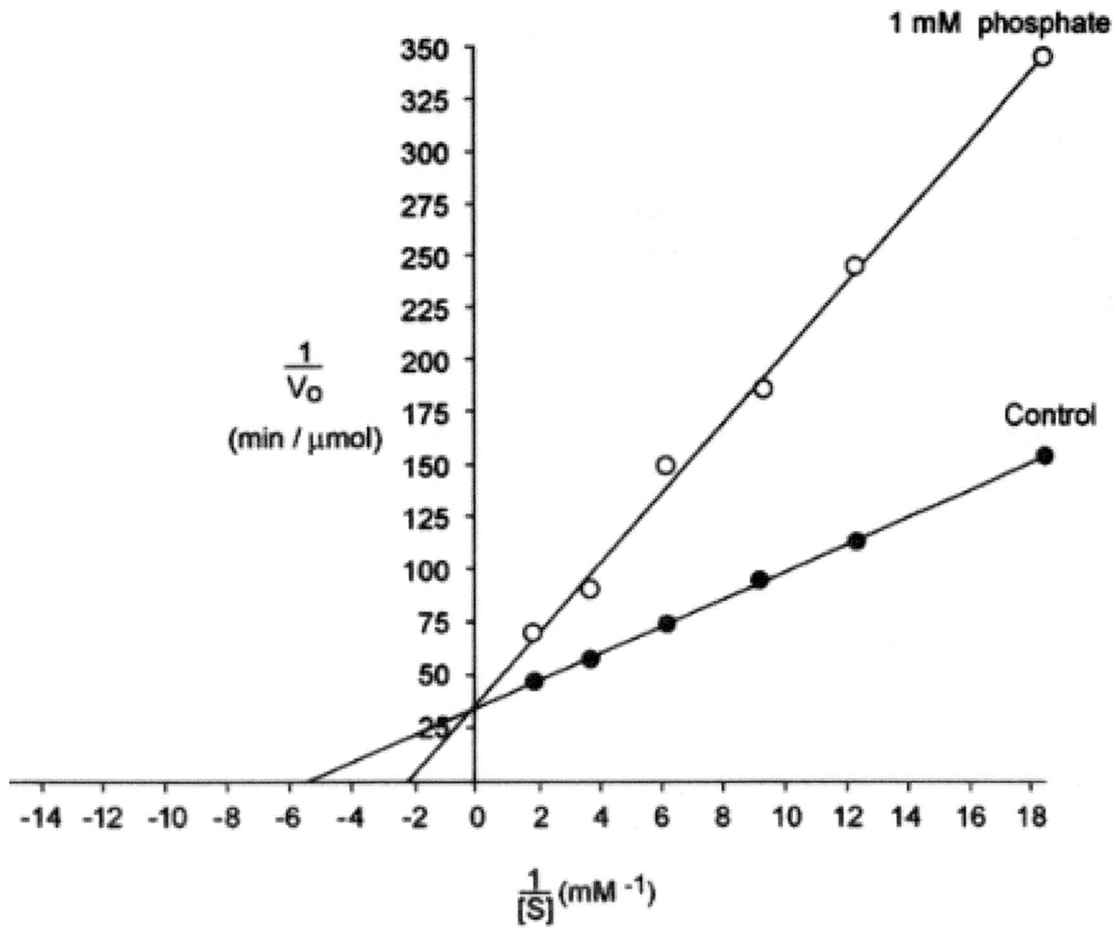


Source for graphs:

Dean, R. L. (2002). Kinetic Studies with Alkaline Phosphatase in the Presence and Absence of Inhibitors and Divalent Cations. Biochemistry and Molecular Biology Education, 30(6), 401-407.

123

**Value: 10pt**
**Late penalty: -5pt per day (i.e. 50% per day)**

The following three questions should be submitted in one "Lab 1B" document to Turnitin by the deadline. **Be <u>certain</u> to include the question number (Q1, Q2, Q3) along with your response, so that each question can be scored separately as it is graded.**

Q1. **Description of the ALPI gene (3pt).** Write a paragraph that addresses the following for ALPI in complete sentences, using your own words:
- The gene's function
- What chromosome it is located on
- Whether it is on the + or – strand
- How long the gene is in bp
- How many exons it has
- Whether it contains any EcoRI cut sites
- What cell type(s) the mRNA is most highly expressed in
- The expected length of the protein
- The names of three vertebrate species that are included in the 'Comparison' track – what caught your eye about these three?

Q2. **Primer selection and visualization (3pt).** The response to this question should contain one item:
- A photo that shows the location of your annealed primers after zooming out 10x, with your photo ID in the lower right of the screen.

Q3. **View and description of your chosen gene (4pt).** The response to this question should contain two items:
a. A photo that shows the additional gene that you chose (<u>not</u> ALPI, TP53, or ACTB) in the genome browser, after zooming out 3x, with your photo ID in the lower right of the screen.
b. A brief description of this gene in complete sentences, stating the gene symbol and gene name, the function, the number of amino acids in the protein, the cell types in which this gene is most highly expressed, and what search term(s) or process you used in order to find this gene in the genome. Be sure to be clear with the last item – help us to understand how you chose this particular gene in the first place!

**Value: 10pt**
**Late penalty: -5pt per day (i.e. 50% per day)**

The following questions should be submitted in one "Lab 2B" document to Turnitin by the deadline. **Be <u>certain</u> to include the question number (Q1, Q2, Q3) along with your response, so that each question can be scored separately as it is graded.**

Q1.   **Exploring PDB files and secondary structure (Jmol Exercise 1).** The response to this question should contain two components **(2pt each)**:

    a.   **One photo** captured at the indicated spot in the instructions manual from Exercise 1 (step xxxiv). This should include the molecule window, the Script Console window, and in the bottom right should include your student ID card.

    b.   **One paragraph** reflecting on whether the exercise was successful in having you more vividly appreciate the distinctive order that allows secondary structure to form, compared with lecture coverage of the figures in Section 2.3 (Secondary Structure) of our textbook. Try to pick one portion of the exercise that was particularly helpful in improving your understanding of how secondary structure is arranged or stabilized (or where the R-groups are with respect to the backbone), and use specific wording to say what you were doing in the exercise that was helpful. If you found the exercise <u>un</u>helpful, then provide specific feedback regarding what you think an ideal exercise would contain, if its goal was to provide 3D enhancement to lecture coverage of secondary structure. (If you found part of the exercise confusing, too brief to be useful, etc., then please use this opportunity to convey that feedback.)

Q2.   **Exploring tertiary structure and disulfide bonds (Jmol Exercise 2).** The response to this question should contain two components **(2pt each)**:

    a.   **One photo** captured at the indicated spot in the instructions manual from Exercise 2 (final step). This should include the molecule window, the Script Console window, and in the bottom right should include your student ID card.

    b.   **One paragraph** describing what can be observed using the 'Slab' command with lysozyme (final steps of Exercise 2). Write a paragraph that describes what you saw when using this new command. This part of the exercise had a specific purpose – think about how the residues (amino acids) were color-coded and what you could observe when viewing the protein's exterior compared with when using the 'Slab' command – be sure to specifically describe the color-coding, what it represented, and the overall spatial distribution pattern that could be observed for lysozyme on the surface vs. in the 'Slab' view. Also mention whether you could view empty (unoccupied) space in the 'Slab' view or whether the interior was tightly packed with atoms.

**Q3.** **Using the ProtParam tool (2pt).** The response to this question should consist of **one paragraph** that addresses the following in complete sentences:

- Lysozyme's molecular weight (kD), most-represented residue (or indication that they are all equally represented), and pI.

- Histone H3's pI, composition in terms of acidic and highly basic residues (and the direct relevance of this to its pI!), and a restatement of the definition of pI (textbook section 3.1).

- An overall comment (your own judgment) regarding whether these two examples suggest that protein composition is largely evenly distributed (most amino acids represented at approximately similar frequency regardless of which protein you were to study) or not.

**Value: 10pt**
**Late penalty: -5pt per day (i.e. 50% per day)**

The following questions should be submitted in one "Lab 3B" document to Turnitin by the deadline. **Be <u>certain</u> to include the question number (Q1, Q2) along with your response, so that each question can be scored separately as it is graded.**

Q1.  **Recreating your own version of textbook Fig. 30.11 (threonyl-tRNA synthetase bound to tRNA and AMP).** The response to this question should contain two components:

   a.  **One high-quality JPEG (2pt)** exported from the Jmol program (not a screenshot!). If you can't choose one angle that best captures what you want to show, then you should export two very different views (most likely one that clearly shows the anticodon and the other that very clearly shows the AMP near the 3' end of the tRNA), and label them 'Figure A' and 'Figure B' in your caption.

   b.  **A clear and very detailed caption (4pt)** that is customized to your own personal image(s) that you created – it should definitely contain differences compared with what your classmates are submitting! There are various orders in which you could present the following information, but at some point, with complete sentences, you should include all of the following:

   - The name of the enzyme and what it is bound to (the tRNA, small molecule, and the cofactor), along with the specific colors used for each.
   - The filename that you downloaded.
   - The name of the program that you used to create the image(s).
   - The specific names of the colors that you used for everything (e.g. 'the majority of the tRNA is colored T, while the individual bases in the anticodon were highlighted as follows:…').
   - The styles that you chose for everything (e.g. 'the AMP molecule is displayed in the spacefill style…').
   - The organism, the resolution in Angstroms, and the method by which the structure was obtained (X-ray diffraction? NMR? Other?) Remember, you can easily get this information from the appropriate page at http://www.rcsb.org/pdb
   - Something that you want to draw the viewer's attention to. (Notice how close together two particular things are…? Notice how large or how small something is…? Notice how much or how little secondary structure something has…? Notice how the color-coding reveals a

particular thing that would otherwise have been hard to see…?) Clearly, aside from your color choices, this is one of your best chances to put your personal stamp on your submission. Ideally you can find something to point out that was somewhat hard to discern in the original Fig. 30.11!

**Q2.** **The ribosome (Jmol Exercise 19).** The response to this question should contain two components **(2pt each)**:

a. **One photo** captured at the indicated spot (final step) in the instructions from Exercise 19. This should include the molecule window, the Script Console window, and in the bottom right should include your student ID card. The purpose is to, as best you can, show that where the two tRNAs are close together in the peptidyl transferase center (which of course is the opposite end from where the mRNA is!) there is predominantly **rRNA**, not ribosomal protein.

b. **One paragraph** reflecting on this Ribosome exercise. There were three separate files to upload because the final 'intact' ribosome file lacked many of the ribosomal proteins and did not allow as good of an investigation of the RNA vs. protein localization as the first two files did. Was this three-file exercise effective in its intended purpose of having you more deeply understand the embedded content in textbook Figs. 30.14 (showing that the ribosome contains both RNA and protein) and 30.16 (showing tRNA binding to the mRNA and the ribosome, and showing that the peptidyl transferase center is lined with rRNA, causing the ribosome to be described as a 'ribozyme' rather than as an 'enzyme')? If not, then do you have suggestions for how these messages could be better conveyed?

- (Alternatively, if you thought that Exercise 19 went fine, you can instead give a detailed response to the Central Dogma animations. Were they valuable? Did they serve their intended purpose of helping you to visualize the complexity and speed of transcription and translation better? Or do you have suggestions for what types of animations would be more useful based on your impression of what Central Dogma-related processes remain the most difficult to visualize?)

**Value: 10pt**
**Late penalty: -5pt per day (i.e. 50% per day)**

The following questions should be submitted in one "Lab 4B" document to Turnitin by the deadline. **Be <u>certain</u> to include the question number (Q1, Q2, Q3) along with your response, so that each question can be scored separately as it is graded.**

**Q1.**   **3D view of the chymotrypsin inhibitor in Jmol.** The response to this question should contain two components:
   a.   **Photo of 3D inhibitor structure (1pt).** One photo of your screen, captured at the indicated spot in instructions. Both the phenyl group and the peptide bond should be visible, and your student ID card should be visible in the lower-right corner of the photo.
   b.   **Caption for inhibitor structure (2pt).** One brief paragraph to caption your photo. There are various orders in which you could present the following information, but at some point, with complete sentences, you should include all of the following:
   - What filename you downloaded.
   - Where you downloaded it from.
   - That the molecule is being displayed in Jmol.
   - The full name of the molecule (not just 'APF') – refer back to the page where you downloaded the file.
   - The notable structural features (key functional groups and important types of bonds) that can be seen in the actual figure that you are submitting. You are captioning the photo that you're submitting – so don't mention any part of the molecule's structure that isn't actually visible! Your purpose is to orient the reader to what they are seeing in the photo that you submitted.

**Q2.**   **Color-coded views of chymotrypsin and chymotrypsinogen.** The response to this question should contain three components:
   a.   **Photo of chymotrypsin (2pt).** Your color-coded photo of chymotrypsin, with your student ID showing in the lower right.
   b.   **Photo of chymotrypsinogen (2pt).** Your color-coded photo of chymotrypsinogen (using the same color choices), with your student ID showing in the lower right.
   c.   **Caption for chymotrypsin and chymotrypsinogen images (3pt).** A clear and detailed caption that is customized to your own personal images that you created. There are various orders in which you could present the following information, but at some point, with complete sentences, you should include all of the following:
   - An indication that the first image is chymotrypsin (give the filename and state that it is bound to the inhibitor that you showed in the

previous question) and that the second image is its inactive precursor chymotrypsinogen (give its filename).

- The name of the program that you used to create the images.
- The specific names of the colors that you used for everything (e.g. 'the majority of the protein is colored U, while the individual residues of the catalytic triad were highlighted as follows:...').
- The organism, the resolution in Angstroms, and the method by which each structure was obtained (X-ray diffraction? NMR? Other?) Remember, you can easily get this information (and _so much_ more!) by searching http://www.rcsb.org/pdb using the PDB identifiers that you typed in when you used the **File -> Get PDB** command in Jmol to access the structures.
- Something that you want to draw the viewer's attention to. (Notice how close together or far apart two particular things are? Notice how the color-coding reveals a particular thing that would otherwise have been hard to see?) Clearly, aside from your color choices, this is one of your best chances to put your personal stamp on your submission. Try to focus on a comparison between the two images – given that one is active chymotrypsin and one is inactive precursor, what can you point out that looks different? Try your best!

**Value: 10pt**
**Late penalty: -5pt per day (i.e. 50% per day)**

The following questions should be submitted in one "Lab 5B" document to Turnitin by the deadline. **Be <u>certain</u> to include the question number (Q1, Q2, Q3...) along with your response, so that each question can be scored separately as it is graded**.

**Q1.**   **Photo of ibuprofen-bound albumin in Jmol.** The response to this question should contain two components:
   a.   One photo of your screen, captured at the indicated spot in the instructions. A bound molecule should be visible, and your student ID card should be visible in the lower-right corner of the photo.
   b.   One brief paragraph to caption your photo. There are various orders in which you could present the following information, but at some point, with complete sentences, you should include all of the following:
   - what filename you accessed
   - that the structure is being displayed in Jmol
   - the name of the protein and the bound molecule
   - the number of disulfide bonds per monomer, and whether any of them link the monomers together

**Q2.**   **Photo of alkaline phosphatase in Jmol.** The response to this question should contain two components:
   a.   One photo of your screen, captured at the indicated spot in the instructions. A molecule of bound inhibitor and at least one glycosylation site should be visible, and your student ID card should be visible in the lower-right corner of the photo.
   b.   One brief paragraph to caption your photo. There are various orders in which you could present the following information, but at some point, with complete sentences, you should include all of the following:
   - what filename you accessed
   - that the structure is being displayed in Jmol
   - the name of the enzyme, the bound molecule, the metals, and the name of the carbohydrate
   - the number of disulfide bonds per monomer, and whether any of them link the monomers together

**Q3.**   **Photo of chymotrypsin in Jmol.** The response to this question should contain two components:
   a.   One photo of your screen, captured at the indicated spot in the instructions. A molecule of bound inhibitor should be visible, and your student ID card should be visible in the lower-right corner of the photo.

b.  One brief paragraph to caption your photo. There are various orders in which you could present the following information, but at some point, with complete sentences, you should include all of the following:
- what filename you accessed
- that the structure is being displayed in Jmol
- the name of the enzyme and the bound molecule
- the fact that one functional copy of the enzyme contains three chains, and whether disulfide bonds link any of the chains together

**Q4.  Photo of malate dehydrogenase in Jmol.** The response to this question should contain two components:

a.  One photo of your screen, captured at the indicated spot in instructions. A molecule of bound substrate and electron carrier should be visible (perhaps overlapping and obscuring each other though – this is fine), and your student ID card should be visible in the lower-right corner of the photo.

b.  One brief paragraph to caption your photo. There are various orders in which you could present the following information, but at some point, with complete sentences, you should include all of the following:
- what filename you accessed
- that the structure is being displayed in Jmol
- the name of the enzyme and the two bound molecules
- whether any disulfide linkages are present

**Q5.  Photo only of malate and NAD+ (malate dehydrogenase hidden) in Jmol.** The response to this question should contain two components:

a.  One photo of your screen, captured at the indicated spot in instructions. Malate should be shown in close proximity to the nicotinamide ring of NAD+, and your student ID card should be visible in the lower-right corner of the photo.

b.  One brief paragraph to caption your photo. There are various orders in which you could present the following information, but at some point, with complete sentences, you should include all of the following:
- that this is again the same structure file as for the previous question, but with the enzyme hidden
- the names of the two molecules, emphasis that the enzyme catalyzes a redox reaction, and the names of the two products that will be formed by the enzyme

# Appendices

The following pages contain the introductory Jmol assignment (downloading Jmol and viewing DNA structure), as well as some general information about the amino acids and a list of all the possible colors that can be used in Jmol.

This assignment should be completed for the lecture portion of the course before proceeding with the lab assignments that use Jmol. It was modified from **EXERCISE 17: DNA** from "**Studying Protein and Nucleic Acid Structure with Jmol. For use with Berg, Tymoczko, Gatto, and Stryer, Biochemistry, 8th Edition. By Jeffrey A. Cohlberg. Revised May 2015**".

1. **Download Jmol**
   - If possible, use a laptop that you can bring to lab, since we will be using Jmol (as well as Excel) in Lab 4A.
   - Running Jmol will require Java version 1.4 or later, so if necessary go to http://www.java.com/ to download this before proceeding with the Jmol download.
   - Now download Jmol. Go to http://jmol.sourceforge.net/ and in the upper right, click on 'Download'.
   - Then under 'Downloading Jmol' click the 'download link'.
   - The download will be in a compressed .zip format – either it will expand automatically or else find the downloaded file and double-click to expand it.
   - Jmol is a 'Java applet' and does not need to be installed. You will open Jmol by looking for the '**Jar**' (=Java archive) file.
     - On a Mac, you can run Jmol by opening the Jmol folder that you expanded and double-clicking on the file called **Jmol.jar**
     - On a PC, it should also be named **Jmol.jar**; if you view the details for each file type then the one to double-click on will be the '**Executable Jar File**'.
   - Double-click to verify that Jmol opens – it should look like the view shown below. If you receive an error message indicating that you can't open the application since it is from an unknown source, you may need to adjust your security preferences to allow you to open Jmol.
     - Troubleshooting:
       - There are reports that in some PC operating systems when you double-click it appears that nothing happens, and that this can be fixed by right-clicking on the Jmol.jar file and choosing the '**Open Outside**' option.
       - Microsoft Surface (Windows Pro 10): tech services in Speare Hall did the following: downloaded WinRAR (stays free forever, even though there is a prompt that says to pay), then downloaded latest version of Java, then downloaded Jmol. Opened Jmol.jar, and set 'Open with' to open with WinRAR. This needs to be done every time; can be made into an automatic setting.

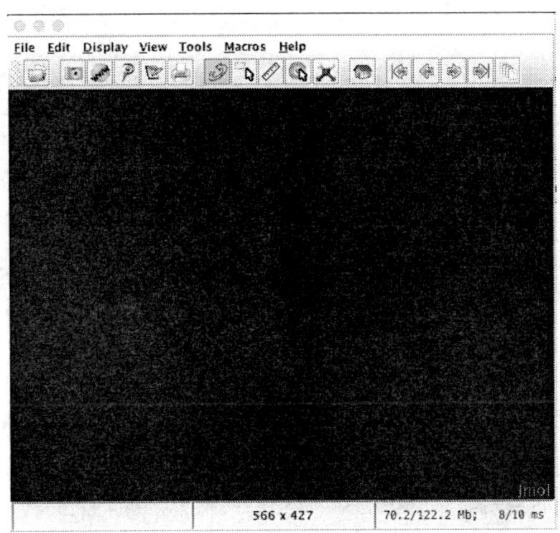

2. **Examine the structure of double-stranded DNA**. You will do a few manipulations of a 'B-form DNA' structure, and then submit one photo of your work. The purposes are (1) to gain familiarity with Jmol well before using it in the lab assignments, and (2) to gain a better appreciation for DNA structure (Ch.1 and Ch.4 in the textbook) through manipulations of a 3D structure.

- Please proceed with the following steps:

    i. In the top menu in Jmol go to **File -> Console** to display the Jmol Script Console. Although Jmol has many options that can be modified using menus, it is often faster and easier to use the console to type in commands. Press the return key after each command that you type.

    ii. Type **set pdbAddHydrogens** into the console (and then press return) – this will ask Jmol to show hydrogens in the structure, even though they could not be resolved in the original structure data due to technical limitations. You should receive the response 'pdbAddHydrogens = true'

    iii. In the top menu in Jmol go to **File -> Get PDB** and enter the identifier **423D**, which is for a standard 'B' form DNA molecule. The file should load fairly quickly; if you receive an error message, then first try again. If you still get an error message, then visit the site http://www.rcsb.org/pdb/ that Jmol is trying to get the file from – if for some reason the website is down temporarily, then you will need to try again a bit later.

    iv. When your DNA file loads, if you rotate it around (click and hold down as you move your cursor), you should immediately note the striking placement of the nitrogenous bases (blue = nitrogen in this standard 'CPK' coloring style, red = oxygen, orange = phosphorus, grey = carbon, white = hydrogen). A useful feature of the console is that it displays information about the file – in this case, it says that there are two molecules (strand A and strand B of the DNA) and gives the sequence. 'D' indicates that the nucleic acid is 'deoxy' (rather than being RNA), and each 'P' indicates a phosphate connecting each nucleoside. If you take a good look at the sequence, you should be able to tell that this sequence has reverse complementarity – two <u>identical</u> sequences are able to bind to each other, and so this is why only one DNA sequence is given rather than two. (If you're not sure why, try writing out what the reverse complement of the given sequence is – you should end up with the same sequence as what you started with! In other words, this sequence binds to itself.)

v. One thing that makes this image a little cluttered is water molecules and magnesium ions that crystallized along with the DNA. Type **delete water or mg** into the console and these will disappear.

vi. Rotate the structure around and try to see the major groove and minor groove. These are very difficult to visualize in the textbook, and so this is one of your best chances, even though this is a short segment of DNA. Imagine a DNA-binding protein approaching the DNA – one that is going to specifically contact the nitrogenous bases. Most DNA-binding proteins bind in the major groove, where there is a lot of space (and therefore the protein can be relatively wide); notice that in the minor groove there is very little space and only an exceptionally skinny protein would be able to enter and contact the information-containing (i.e. nitrogenous base) portion of the DNA.

vii. Before moving on, take a good look at the planarity of the nitrogenous bases, and how the carbohydrates (deoxyriboses) are absolutely not coplanar with the bases. This makes sense, because if the bases are the horizontal 'rungs of the ladder' then the sugar-phosphate backbones are the vertical supports of the ladder. Try to look for examples of the 3' and 5' carbon in deoxyribose (recalling that the 1' carbon is the one attached to the base; the numbering is reviewed in textbook figure 4.2).

viii. Now change the view by typing the command **select backbone; spacefill–** then rotate the structure around and you should see that this highlights the major vs. minor groove even more.

ix. Next we'll put the base pairs in spacefill instead. Type **select dna; spacefill** then **select backbone; spacefill off; wireframe 80** – you should now see a view that is similar in message to textbook Fig. 1.14, which was that the base pairs in double-stranded DNA stack at the optimal van der Waals contact distance, helping to stabilize the DNA structure.

x. Now look at a single base pair by typing **restrict 8 or 17** – you can see that in spacefill view that the two bases in this pair are so close together that the pair appears to be a single molecule.

xi. Type **select 8 or 17; spacefill off; wireframe 80** – now you should be able to see the separation between the bases. (You can use the '**undo**' button on the console to go back to the spacefill view to compare, and then the '**redo**' button to go forward again to the wireframe view.) Again, if you rotate this structure you should see clearly that the deoxyriboses are not remotely in the same plane as the nitrogenous bases. Further, if you rotate carefully, you should be able to see that each deoxyribose is not fully planar itself – this is also illustrated in textbook Fig. 4.15.

xii. Now it's time to take a picture of a DNA view for your assignment submission. Type **restrict all; select all** and then type in a style (**spacefill** or **cartoon only** or **wireframe 80 only** or **wireframe {some other number} only** or **trace only)**. Or, if you're feeling creative, experiment more with the pop-up menu to try out some other 'Style' and/or 'Color' and/or 'Surfaces' options. Choose a background color from the very many options available at [http://jmol.sourceforge.net/jscolors/#JavaScript%20colors](http://jmol.sourceforge.net/jscolors/#JavaScript%20colors) (and listed in the appendix of the lab manual), and once you have chosen a background color that has name 'X' then change the background with the command **background X** (e.g. changing it to black would have been **background black**).

xiii. Rotate your DNA molecule to a pleasing angle, and take a photo of your screen that shows three things: your DNA on your chosen background, your console, and your student ID in the lower right. Upload this photo for your response to the 'Introduction to Jmol – DNA' assignment for the lecture portion of the course. Some of these artful creations may be shown in lecture! (Note: it's not important that your submission photo be a beautiful high quality image – we'll work on exporting high-quality Jmol images later. Rather, your submission should be a photo that shows three things in one image, like the one below: (1) your molecule, (2) the console so that at least some of the commands that you typed in are visible, and (3) the student ID card (not just a typed in ID number) of you as the person who did the work for this assignment. Don't just make a screenshot on your computer – please do the extra step of taking a photo that shows your ID card, which is what you'll also be doing for the other dry labs.

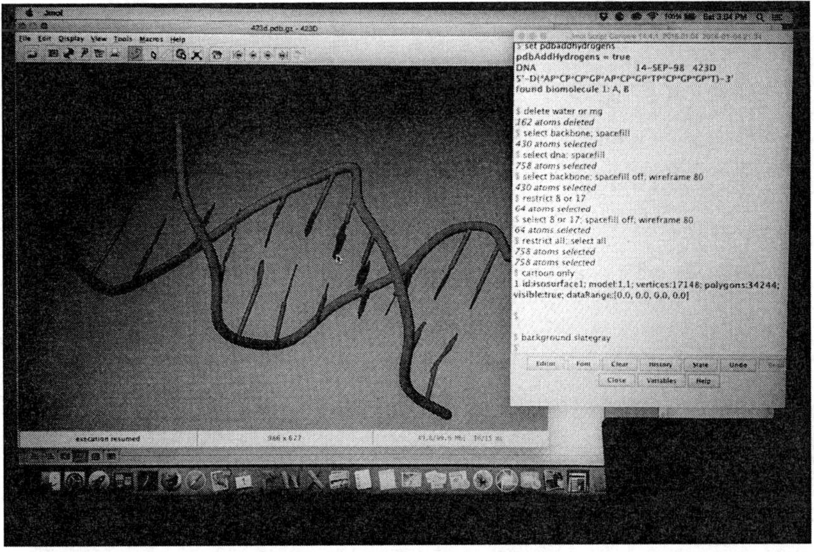

There are 140 colors available for use in customizing structures in Jmol. Of course, it's recommended that you look at the actual color samples at http://jmol.sourceforge.net/jscolors/#JavaScript%20colors

For your convenience, however, they are given here as well:

| | | | | |
|---|---|---|---|---|
| aliceblue | darkolivegreen | indigo | mediumpurple | purple |
| antiquewhite | darkorange | ivory | mediumseagreen | red |
| aqua | darkorchid | khaki | mediumslateblue | rosybrown |
| aquamarine | darkred | lavender | mediumspringgreen | royalblue |
| azure | darksalmon | lavenderblush | mediumturquoise | saddlebrown |
| beige | darkseagreen | lawngreen | mediumvioletred | salmon |
| bisque | darkslateblue | lemonchiffon | midnightblue | sandybrown |
| black | darkslategray | lightblue | mintcream | seagreen |
| blanchedalmond | darkturquoise | lightcoral | mistyrose | seashell |
| blue | darkviolet | lightcyan | moccasin | sienna |
| blueviolet | deeppink | lightgoldenrodyellow | navajowhite | silver |
| brown | deepskyblue | lightgreen | navy | skyblue |
| burlywood | dimgray | lightgrey | oldlace | slateblue |
| cadetblue | dodgerblue | lightpink | olive | slategray |
| chartreuse | firebrick | lightsalmon | olivedrab | snow |
| chocolate | floralwhite | lightseagreen | orange | springgreen |
| coral | forestgreen | lightskyblue | orangered | steelblue |
| cornflowerblue | fuchsia | lightslategray | orchid | tan |
| cornsilk | gainsboro | lightsteelblue | palegoldenrod | teal |
| crimson | ghostwhite | lightyellow | palegreen | thistle |
| cyan | gold | lime | paleturquoise | tomato |
| darkblue | goldenrod | limegreen | palevioletred | turquoise |
| darkcyan | gray | linen | papayawhip | violet |
| darkgoldenrod | green | magenta | peachpuff | wheat |
| darkgray | greenyellow | maroon | peru | white |
| darkgreen | honeydew | mediumaquamarine | pink | whitesmoke |
| darkkhaki | hotpink | mediumblue | plum | yellow |
| darkmagenta | indianred | mediumorchid | powderblue | yellowgreen |

Approximate pKa values are given, based on information in our textbook, *Biochemistry*, 8th Edition, by Berg, Tymoczko, Gatto, and Stryer, 2015.

| name | 1-letter code | 3-letter code | classification | pKa |
|---|---|---|---|---|
| glycine | G | Gly | hydrophobic | none given |
| alanine | A | Ala | hydrophobic | none given |
| proline | P | Pro | hydrophobic | none given |
| valine | V | Val | hydrophobic | none given |
| leucine | L | Leu | hydrophobic | none given |
| isoleucine | I | Ile | hydrophobic | none given |
| methionine | M | Met | hydrophobic | none given |
| tryptophan | W | Trp | hydrophobic | none given |
| phenylalanine | F | Phe | hydrophobic | none given |
| serine | S | Ser | polar | none given |
| threonine | T | Thr | polar | none given |
| tyrosine | Y | Tyr | polar | 11 |
| asparagine | N | Asn | polar | none given |
| glutamine | Q | Gln | polar | none given |
| cysteine | C | Cys | polar | 8 |
| aspartate | D | Asp | acidic | 4 |
| glutamate | E | Asp | acidic | 4 |
| lysine | K | Lys | basic | 11 |
| arginine | R | Arg | basic | 12 |
| histidine | H | His | basic | 6 |